Explorations in Meteorology

A Lab Manual

Dr. Kenneth C. Crawford
Regents' Professor
School of Meteorology and
Oklahoma Climatological Survey
University of Oklahoma

Dr. Kevin A. Kloesel
Director of Outreach and
Adjunct Associate Professor of Meteorology
Oklahoma Climatological Survey

Dr. Renee A. McPherson
Acting Director and
Adjunct Assistant Professor of Meteorology
Oklahoma Climatological Survey

Andrea Dawn Melvin
Program Manager of K-12 Outreach
Oklahoma Climatological Survey

Dale A. Morris
Program Manager of Public Safety Outreach
Oklahoma Climatological Survey

BROOKS/COLE
CENGAGE Learning™

Australia • Brazil • Japan • Korea • Mexico • Singapore • Spain • United Kingdom • United States

BROOKS/COLE
CENGAGE Learning™

Explorations in Meteorology: A Lab Manual
Oklahoma Climatological Survey

Earth Science Editor: Keith Dodson

Assistant Editor: Carol Benedict

Editorial Assistant: Megan Asmus

Technology Project Manager:
  Ericka Yeoman-Saler

Marketing Manager: Jennifer Somerville

Marketing Assistant: Michele Colella

Advertising Project Manager: Kelley McAllister

Project Manager, Editorial Production:
  Hal Humphrey

Print Buyer: Barbara Britton

Permissions Editor: Kiely Sisk

Text Designer: John Humphrey

Copy Editor: Renee McPherson, Dale Morris,
  Andrea Melvin

Illustrator: Ryan Davis, John Humphrey

Cover Designer: Lisa Henry

Cover Image: Ryan Davis

Compositor: Stdrovia Blackburn

For product information and technology assistance, contact us at
Cengage Learning Customer & Sales Support, 1-800-354-9706
For permission to use material from this text or product,
submit all requests online at **www.cengage.com/permissions**
Further permissions questions can be emailed to
permissionrequest@cengage.com

ISBN-13: 978-0-495-01029-6

ISBN-10: 0-495-01029-4

Brooks/Cole
20 Davis Drive
Belmont, CA 94002
USA

Cengage Learning is a leading provider of customized learning solutions with office locations around the globe, including Singapore, the United Kingdom, Australia, Mexico, Brazil, and Japan. Locate your local office at **www.cengage.com/global**

Cengage Learning products are represented in Canada by Nelson Education, Ltd.

To learn more about Brooks/Cole, visit **www.cengage.com/brookscole**

Purchase any of our products at your local college store or at our preferred online store **www.cengagebrain.com**

Printed in the United States of America
7 8 9 10 11 12 13 21 20 19 18 17

DEDICATION

This laboratory manual is dedicated to the educators, students, and public safety officials who originally motivated the Oklahoma Climatological Survey (OCS) to create lessons and case studies. These officials, educators, and students helped OCS to refine and improve a number of tools and resources that serve as a foundation for this manual.

Since 1992, K-12 teachers and students have worked with OCS to design lessons that apply data from the Oklahoma Mesonet (http://www.mesonet.org/) and other sources in their classroom curriculum. Using real-time data, custom display software, and lessons available at http://earthstorm.ocs.ou.edu/, students can learn science and mathematics through classroom activities, science fair projects, and service to their local communities.

Since 1996, public safety officials have applied real-time data from the Oklahoma Mesonet, National Weather Service Doppler radars, and other sources to make life-saving and cost-reducing decisions for their local communities. Using resources available at http://okfirst.ocs.ou.edu/, hundreds of emergency management, law enforcement, and fire service officials across Oklahoma protect their communities with current weather information and forecasts. OCS provides continuing education to these officials to enable them to make better decisions during both hazardous and benign weather.

Finally, we also dedicate this laboratory manual to Mr. David J. Shellberg, our colleague and friend. An enthusiastic supporter of the Oklahoma Mesonet, David worked alongside OCS customers, including K-12 teachers and students, to demonstrate the value of weather data in daily activities. David was killed tragically on October 31, 1994, on American Eagle Flight 4184 from Indianapolis, IN to Chicago, IL.

PREFACE

As both a state agency and a research unit of the University of Oklahoma's (OU) College of Geosciences, the Oklahoma Climatological Survey (OCS) is dedicated to acquiring, processing, archiving, and disseminating climate and weather data and to producing information beneficial for decision-makers and citizens of the State of Oklahoma. The Oklahoma State Legislature established OCS in 1980 to serve as the office of the State Climatologist. The Survey maintains an extensive array of climatological information, operates the Oklahoma Mesonet, and hosts a wide variety of educational outreach and scientific research projects. More information about OCS is available at http://www.ocs.ou.edu/.

In support of its mission, OCS developed two software packages, known as WeatherScope and WxScope Plugin. Together they serve as the primary display software packages for maps and graphs in this manual. WxScope Plugin and WeatherScope have served as foundations for the operational programs (e.g., the Oklahoma Mesonet), outreach services (e.g., for educators, agriculturist, and public safety officials), and research projects of OCS. In particular, WeatherScope, OCS's newest visualization software for current Windows and Mac operating systems, includes capabilities to re-project data from different map projections "on-the-fly" and displays both real-time and archived weather information. The software is available freely for non-commercial use at http://sdg.ocs.ou.edu/.

This publication would not have been created without tremendous contributions from a host of talented professionals. Stdrovia Blackburn (OCS Visual Communications Specialist), Ryan Davis (OCS Student Graphic Designer), and John Humphrey (OCS Graphic Designer) designed the layout and graphics for the lab manual. We greatly appreciate their continuous zeal to excel and their ability to juggle deadlines without complaint and with their usual excellence.

Dr. Andy White, Associate Director of the OU School of Meteorology, helped initiate the creation of the laboratory manual when he and Dr. Ken Crawford were faced with teaching introductory meteorology classes with an outdated manual. During the first year of the manual's development, Dr. White met weekly with teaching assistants, critiqued the content of the evolving manual, and interacted with students to ensure a successful beginning of the lab manual through his support during several introductory courses.

Jennifer Adams, Graduate Teaching Assistant in the OU School of Meteorology, provided comprehensive reviews of the early documents, actively sought student feedback for several years, and was proactive at improving the lab manual. Ms. Cerry Leffer, OCS's Administrative Manager and Public Relations Specialist, facilitated the contractual agreements between OCS, the University of Oklahoma, and the publisher in a manner that ultimately will serve the customers of OCS data and products. Mark Shafer, OCS's Director of Climate Information and Public Policy Specialist, quickly developed one of the 16 lab exercises even though he was not originally responsible to complete the lab exercise.

The authors also thank the following people for their help developing, producing, and testing the materials in this laboratory manual: Derek Arndt, Christy Carlson, Beth Clarke, Tim Decker, John Ensworth, Lori Garcia, Sally Garrison, Loren Gmachl, Donald Giuiliano, Elaine Godfrey, Kevin Goebbert, Dr. Petra Kastner-Klein, Dr. Mark Laufersweiler, Craig Lengyel, Adrian Loftus, Adam Lopes, Gary McManus, Regina McNabb, Billy McPherson, Andrew Mercer, Harold Peterson, Dr. Michael Richman, and Anneliese Sherer. In addition, we appreciate the candid and thorough reviews conducted during the introductory meteorology classes at the University of Oklahoma School of Meteorology.

Primary funding for the development of this manual was through the Oklahoma Climatological Survey, funded through the University of Oklahoma by the Oklahoma State Regents for Higher Education. The School of Meteorology of the University of Oklahoma provided additional funding. The following organizations provided data and products: the Oklahoma Mesonet (a joint program of the University of Oklahoma and Oklahoma State University); the National Oceanic and Atmospheric Administration (through the National Weather Service and the National Environmental Satellite, Data, and Information Service); and the Atmospheric Radiation Measurement Program (a research facility of the U.S. Department of Energy).

TABLE OF CONTENTS

North American Geography

LAB ACTIVITY OBJECTIVES:

- Given a map outline of North America and its country boundaries, you will label the locations of: the United States, Canada, and Mexico; the Atlantic and Pacific Oceans; the Gulf of Mexico; and the Great Lakes.
- Given a map outline of the United States and Canada and their state/province/territory boundaries, and given a list of U.S. states and Canadian provinces, you will label the locations of all states and provinces.
- Given a map outline of the Great Lakes, you will label the locations of all five lakes.
- Given a time and U.S. time zone, you will compute the time in UTC (Universal Coordinated Time). Given a time in UTC, you will compute the time in various U.S. time zones.

MATERIALS NEEDED:

- Laboratory manual
- Pencil and eraser
- Atlas (optional)

GLOSSARY:

Air Mass

Atmospheric Instability

Contiguous

Continental Polar Air Mass

Continental Tropical Air Mass

Jet Stream

Maritime Polar Air Mass

Maritime Tropical Air Mass

Orographic

Topography

Windward

BACKGROUND:

Physical geography is the study of the location and spatial variation of physical phenomena (such as tornadoes or mountains). An understanding of physical geography is an essential part of our study of meteorology and climatology. Weather and climate patterns often are described geographically. We track the movement of weather systems by referring to states and provinces over which weather features pass. We classify **air masses** by geographical regions where air masses form. Specific geographical regions have their own weather and climate patterns. For example, the "Bermuda High" is a high-pressure system that helps to make the southeast United States (U.S.) warm and humid during the summer months.

Physical geographers sometimes describe the earth in terms of three interrelated geographical systems, called spheres. The *atmosphere* surrounds the earth and is composed of a mixture of gases (primarily nitrogen and oxygen). The *hydrosphere* contains all the water substance on the planet including all surface and subsurface water bodies. The *lithosphere* is the land surface and subsurface including the crust, mantle, outer core, and inner core. In addition, the *biosphere* consists of the regions of these three spheres where living organisms exist. Each sphere interacts with all other spheres and contains portions of the other spheres. For example, a major component of the study of meteorology is *hydrometeorology*, or the study of the impact of water in all its phases upon the atmosphere, including the processes of *evaporation* from the land surface and *transpiration* from vegetation.

Land features such as lakes, mountains, and even hills affect the lower atmosphere, for example, by changing the angle and speed of surface wind flow. Meteorologists recognize that **orographic** lift – the process of air ascending as a result of the presence of a mountain – is responsible for much of the rainfall and snowfall in **windward** areas like the western side of the Rocky Mountains. When we study climatology, we will see that physical geography and climate are linked through seasonal patterns of temperature and precipitation, which differ across mountain ranges.

In this laboratory exercise, you will review the locations of major geographical features of importance to U.S. and Canadian meteorologists, including each state of the U.S. and the provinces and territories of Canada, major mountain ranges and lakes of the U.S. and southern Canada, and oceans adjacent to the North American continent.

If you have difficulty with this laboratory exercise, you should spend part of your study time each week reviewing the maps included with this exercise. To talk fluently with meteorologists, you must know these locations on a map.

WORLD GEOGRAPHY

The earth itself presents an interesting challenge to meteorologists. The atmosphere is a gaseous fluid that is coupled loosely to a rotating sphere. The rotating earth introduces dynamic forces that act upon the atmosphere. The interactions vary by both latitude and proximity to the surface. Some forces are more pronounced near the equator; others are more evident near the poles. Surface friction acts on air near the earth's surface; such effects often disappear in the middle atmosphere.

Latitude bands

In introductory meteorology, we generally divide the earth at the equator into the Northern and Southern Hemispheres (see Figure 1). We further divide each hemisphere into latitude bands. The *tropics* begin at the equator and range in latitudes between the Tropic of Cancer (23.5° N) and the Tropic of Capricorn (23.5° S). Although weakly defined, the *sub-tropics* generally are thought to range from 23.5° N to 30° N and from 23.5° S to 30° S. The *mid-latitudes* span the latitude range from 30° to 60° N and from 30° to 60° S. Most of the United States is in the mid-latitudes. The *polar regions* extend from 60° to 90° N (the North Pole) and from 60° to 90° S (the South Pole).

Figure 1 – Diagram of Latitude Regions

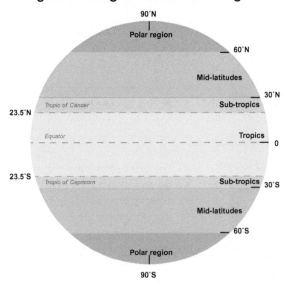

You should learn the locations of the seven continents and the major oceans with respect to these latitude bands.

UNITED STATES

State boundaries

The United States makes up the bulk of the southern half of North America. The U.S. shares a boundary with our neighbor to the north — Canada. Alaska and Hawaii are separated geographically from the **contiguous** states or "lower 48." Review and memorize the location and names of every U.S. state (see Table 1).

Regions

The United States is so large that the country experiences different weather and climate patterns based on location and surface features. For example, the weather and climate of the Rocky Mountains are different from that of the Gulf Coast. It often is useful to divide the United States into regions. Unfortunately, not all regions have clear-cut boundaries. Important regions include New England, the Mid-Atlantic, Great Lakes, Gulf Coast, Southeast, Great Plains, Rocky Mountains, Intermountain West, Great Basin, Pacific Northwest, Southern California, and Southwest.

Mountains

Mountains, and even low hills, influence wind flow in the lower atmosphere. You should learn the location, extent (i.e., length), general height and width, and

Table 1 – Abbreviations of U.S. States and the Canadian Provinces and Territories

The United States of America					
Alabama	AL	Louisiana	LA	Ohio	OH
Alaska	AK	Maine	ME	Oklahoma	OK
Arizona	AZ	Maryland	MD	Oregon	OR
Arkansas	AR	Massachusetts	MA	Pennsylvania	PA
California	CA	Michigan	MI	Rhode Island	RI
Colorado	CO	Minnesota	MN	South Carolina	SC
Connecticut	CT	Mississippi	MS	South Dakota	SD
Delaware	DE	Missouri	MO	Tennessee	TN
Florida	FL	Montana	MT	Texas	TX
Georgia	GA	Nebraska	NE	Utah	UT
Hawaii	HI	Nevada	NV	Vermont	VT
Idaho	ID	New Hampshire	NH	Virginia	VA
Illinois	IL	New Jersey	NJ	Washington	WA
Indiana	IN	New Mexico	NM	West Virginia	WV
Iowa	IA	New York	NY	Wisconsin	WI
Kansas	KS	North Carolina	NC	Wyoming	WY
Kentucky	KY	North Dakota	ND		
The Canadian Provinces and Territories					
Alberta	AB	Newfoundland	NF	Ontario	ON
British Columbia	BC	Northwest Territories	NT	Prince Edward Island	PE
Manitoba	MB	Nova Scotia	NS	Quebec	QC
New Brunswick	NB	Nunavut	NU	Saskatchewan	SK
				Yukon Territory	YT

orientation of the following important mountain ranges: Rocky Mountains, Appalachians, Alleghenies, Catskills, Sierras, Cascades, and Ozarks. The Rocky Mountains are younger, higher, and more jagged than are the Appalachians, which are older, worn down, and rounded. The orientation of all major mountain ranges in the U.S. is north to south; thus, these ranges are perpendicular to the overall west-to-east airflow that normally is observed aloft.

Large bodies of water

The Great Lakes, located near the center of the northern boundary of the United States, are among the world's largest freshwater lakes. They can affect weather conditions profoundly along their shores by supplying moisture for heavy snowfalls. The Gulf of Mexico provides atmospheric moisture to the eastern half of the contiguous U.S. Many storms that reach the Pacific Northwest originate in the Gulf of Alaska. We soon will study the influences of land and water, especially the Pacific Ocean and the Atlantic Ocean. Large bodies of water, whether inland lakes, gulfs, or oceans, influence regional weather and climate profoundly. You should know the location and name of each of these large bodies of water.

CANADA

Weather, climate, and physical geography span political boundaries. The Pacific Coast links Washington with Alaska through beautiful British Columbia. The Rocky Mountains extend northward through the province of Alberta. The Great Plains extend northward through Saskatchewan and Manitoba. North of the Great Lakes and New England, we find Ontario and Québec. The maritime provinces of Newfoundland, New Brunswick, Nova Scotia, and Prince Edward Island share the northern Atlantic coastline with Maine.

The United States–Canada boundary

The United States–Canada border lies at approximately 55° N, just south of the imaginary boundary between the middle latitude and polar regions. We soon will learn that this latitude often marks a region of low pressure and storminess. Many of the storm systems that affect both the United States and Canada originate and often track along this boundary. These storms acquire their energy from the polar **jet stream**, which often is found at high altitudes along the United States–Canadian border.

Air mass source regions

The polar jet stream, or simply polar jet, marks the boundary between cold, often dry, air to the north and warm, often moist, air to the south. In fact, cold air masses that reach the United States often have their origins in Canada. Central and northern Canada — especially Alberta, Saskatchewan, and Manitoba — are the source regions for a particular **air mass** called **continental polar** (cP). The severe weather of the Great Plains is produced when cP air clashes with **maritime tropical** (mT) air from the Gulf of Mexico or the southern Atlantic Ocean. Cold, moist air masses, known as **maritime polar** (mP), flow from the Pacific Ocean west of Oregon, Washington, and British Columbia. Maritime polar air also can originate over the maritime provinces near the Atlantic. Both source regions influence the weather of the northern tier of the United States. A later lab will examine air masses in more detail. For this lab, you should work with the practice maps and learn the locations and names of the Canadian provinces and territories (see Table 1).

MEXICO

The United States–Mexico border slices through the desert Southwest. During the summer, when the sun is near its summer solstice, the high, flat Mexican Plateau absorbs much of the sun's incoming energy and the overlying air becomes hot and dry. The resulting **continental tropical** (cT) air mass can move northeastward from Mexico and across eastern Arizona, New Mexico, and West Texas into the Great Plains. Though it may seem improbable, a cT air mass contributes to **atmospheric instability** and often helps create severe weather in the spring and summer across the southern Great Plains.

TIME CONVERSION

Because meteorological data are observed around the globe, an international time standard was established to avoid confusion and to allow for easy comparison of observations that were measured simultaneously worldwide. The current standard is based upon the time in Greenwich, England, where the Prime Meridian (0° longitude) is located. The Greenwich Mean Time

(GMT) standard uses military time — a 24-hour clock. The naming convention has changed over the years, and now we refer to this same standard as UTC, for Universal Coordinated Time. In addition, you may hear this standard referred to as Zulu time, or simply Zulu or Z. For simplicity, we will be consistent and use UTC.

To convert from local time to UTC, first convert the time to the 24-hour clock used in the military. For example, 6:00 AM converts to 0600 hours; 1:00 PM converts to 1300 hours; 7:00 PM converts to 1900 hours. The conversion step from military time to UTC is based on the number of time zones between the given location and Greenwich, England (at 0° longitude). Because the earth is a sphere and there are 24 hours in a day, each time zone represents approximately 15° longitude (360° longitude/24 hours = 15° longitude per hour). Central Standard Time (CST), the time zone for much of the central U.S. during winter, is 6 hours behind UTC; that is, CST is 6 time zones to the west of Greenwich, England. Thus UTC = CST + 6. Hence, 6:00 AM CST converts to 1200 UTC (i.e., 6:00 AM CST to 0600 hours CST plus 0600 hours equals 1200 hours UTC, or 1200 UTC); 1:00 PM CST converts to 1900 UTC (i.e., 1:00 PM CST to 1300 hours CST plus 0600 hours equals 1900 hours UTC, or 0700 UTC); 7:00 PM converts to 0100 UTC the next day (i.e., 7:00 PM CST to 1900 hours CST plus 0600 hours equals 2500 hours UTC, or 0100 UTC the following day).

To convert from UTC to local time, simply use the equation above (UTC = CST + 6) and solve for CST; so, CST = UTC - 6. For example, 0800 UTC is 2:00 AM CST (0800 UTC minus 0600 hours equals 0200 hours CST, or 2:00 AM).

UTC does not change with Daylight Savings Time (DST), so each location that observes DST moves ahead one hour (e.g., CDT = UTC - 5). Tables 2 and 3 summarize the difference in time between UTC and various time zones across the United States and Canada.

Table 2 – Conversion of U.S. Time Zones to UTC

Eastern Standard Time (EST)	UTC – 5 hours	Eastern Daylight Time (EDT)	UTC – 4 hours
Central Standard Time (CST)	UTC – 6 hours	Central Daylight Time (CDT)	UTC – 5 hours
Mountain Standard Time (MST)	UTC – 7 hours	Mountain Daylight Time (MDT)	UTC – 6 hours
Pacific Standard Time (PST)	UTC – 8 hours	Pacific Daylight Time (PDT)	UTC – 7 hours

Table 3 – Conversion of UTC to U.S. Time Zones

Eastern Standard Time (EST)	UTC + 5 hours	Eastern Daylight Time (EDT)	UTC + 4 hours
Central Standard Time (CST)	UTC + 6 hours	Central Daylight Time (CDT)	UTC + 5 hours
Mountain Standard Time (MST)	UTC + 7 hours	Mountain Daylight Time (MDT)	UTC + 6 hours
Pacific Standard Time (PST)	UTC + 8 hours	Pacific Daylight Time (PDT)	UTC + 7 hours

LABORATORY EXERCISES:

The purpose of this introductory laboratory is to improve your knowledge of names and locations of regions that are important to U.S. meteorologists. If you have difficulty with this laboratory exercise, you should spend part of your study time each week reviewing the maps included with this exercise. To talk fluently with meteorologists, you must know these locations on a map.

Use the maps in this lab exercise to practice locating and naming the political and physical boundaries of North America. Know the locations and names of the following locations:

- The United States, Canada, and Mexico — the countries themselves.
- Each of the 50 United States and the Canadian Provinces.
- The regions of the United States, Canada, and Mexico.
- The major mountain ranges of the United States and southern Canada.
- The Great Lakes, gulfs, and oceans bordering the United States.

In addition, you should know the locations of the world's continents and oceans as well as how to convert between UTC and several local U.S. time zones.

Part I: Locating Geographical Features

1. Using the world map (Figure 2), label the following continents and oceans: Africa, Antarctica, Asia, Australia, Europe, North America, South America, Arctic Ocean, Atlantic Ocean, Indian Ocean, Pacific Ocean, and Southern Ocean.

Figure 2 – World Map (Flat Earth Map Projection)

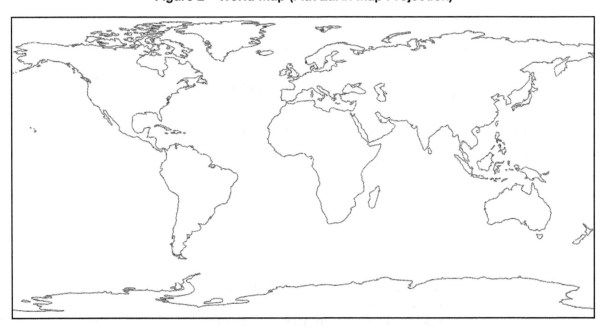

2. Using the North America map (Figure 3), identify the Atlantic and Pacific Oceans, the Gulf of Mexico, and Hudson Bay.

3. Using the North America map (Figure 3), locate and label the Canadian provinces and territories (Table 1). Locate and label the U.S.–Canada and the U.S.–Mexico borders.

Figure 3 – Map of North America (Lambert Conformal Conic Map Projection)

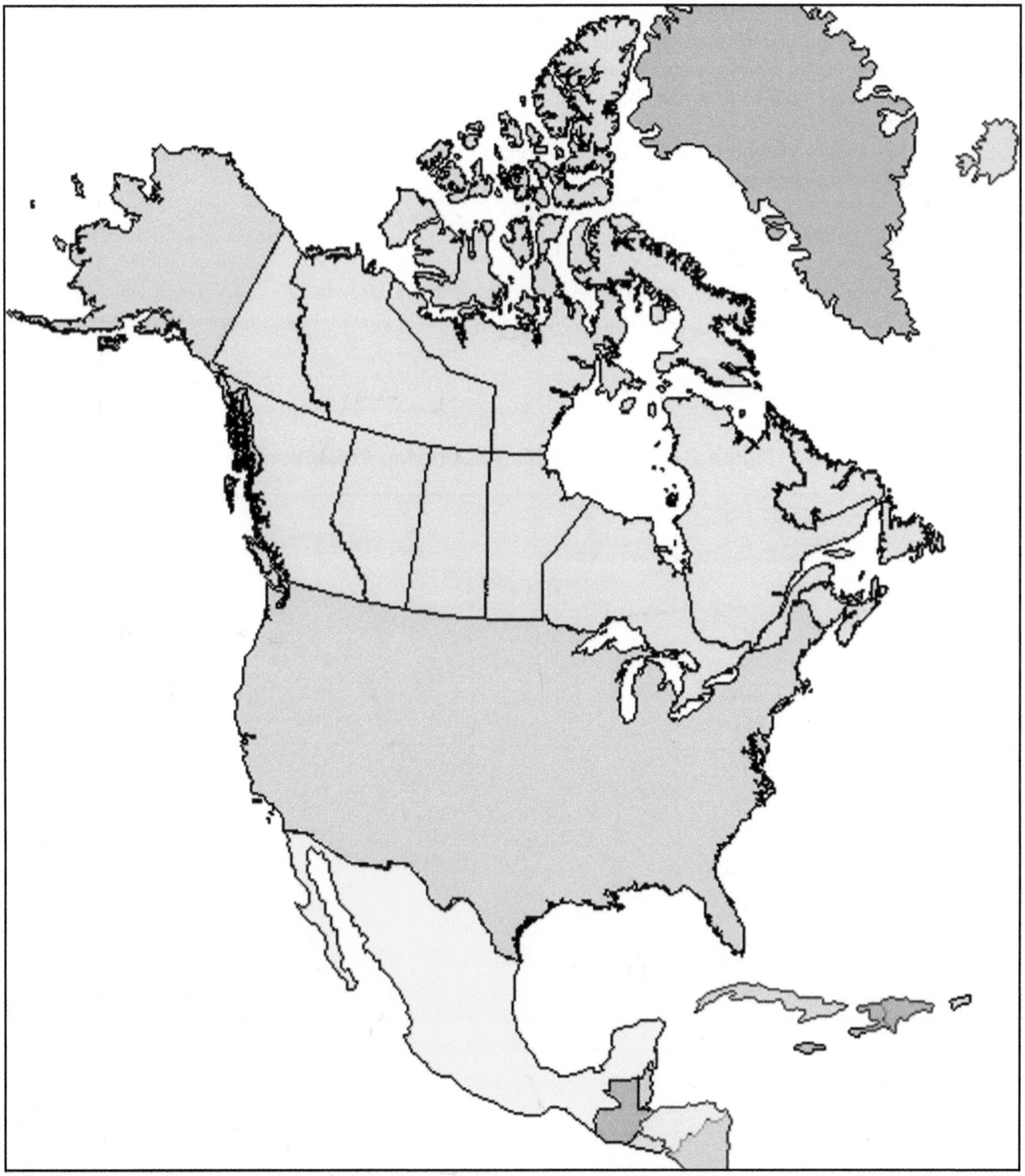

4. Using the United States map (Figure 4) and the list of U.S. states (Table 1), write the two-letter abbreviation for each of the U.S. states.

Figure 4 – Map of United States (Polar Stereographical Map Projection)

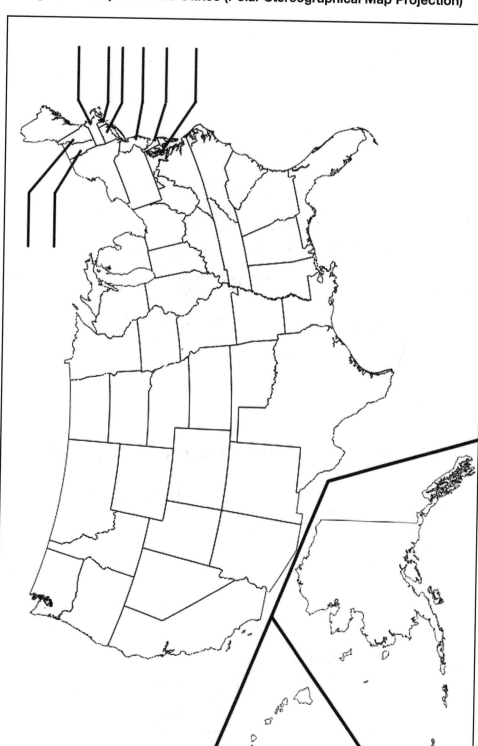

5. Using the map of U.S. **topography** (Figure 5), locate and label the major mountain ranges of the United States (Rocky Mountains; Appalachians; Alleghenies; Catskills; Sierras; Cascades; and Ozarks).

Figure 5 – Topography of the United States

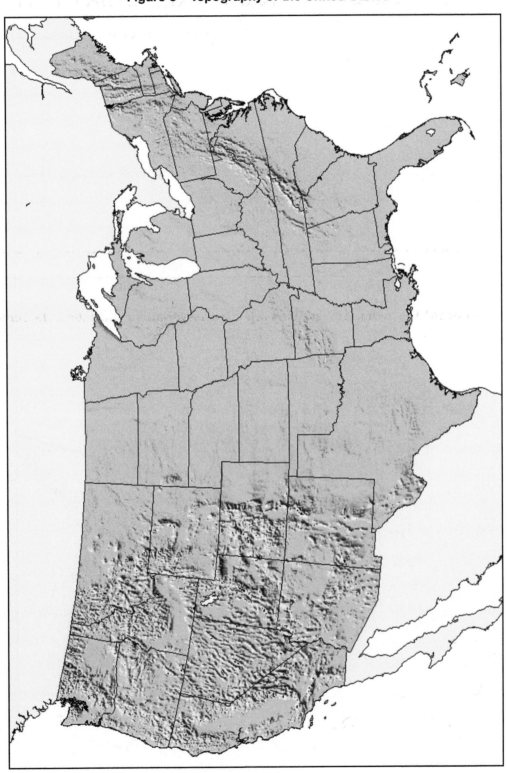

6. Using the map of U.S. topography (Figure 5), locate and label the Great Lakes. (Note: Acronyms — words made from the first letters of a list of words — often help us remember a list of words. An acronym to help remember the Great Lakes is H-O-M-E-S.)

7. Using the U.S. topographic map (Figure 5), draw and label the approximate boundaries for the Great Basin, the Central Great Plains, the Rocky Mountains, the desert Southwest, the Gulf Coast, New England, the Pacific Northwest, the Great Lakes region, the mid-Atlantic states, and the Southeast. Expect some ambiguity and overlap with these regional boundaries. Even physical geography books do not provide unique answers to the location of regions.

PART II: Map Interpretation

8. Using Figure 6, find the four warmest record maximum temperatures. Fill in this table.

State	Warmest Maximum Temperatures

Figure 6 – Record Maximum Temperatures in °F (Courtesy: National Climatic Data Center)

Record Highest
Temperature (°F)
(through 2000)

9. Using Figure 7, find the four coldest record minimum temperatures. Fill in this table.

State	Coldest Minimum Temperatures

10. Using Figure 7, find the four warmest record minimum temperatures. Fill in this table.

State	Warmest Minimum Temperatures

Figure 7 – Record Minimum Temperatures in °F (Courtesy: National Climatic Data Center)

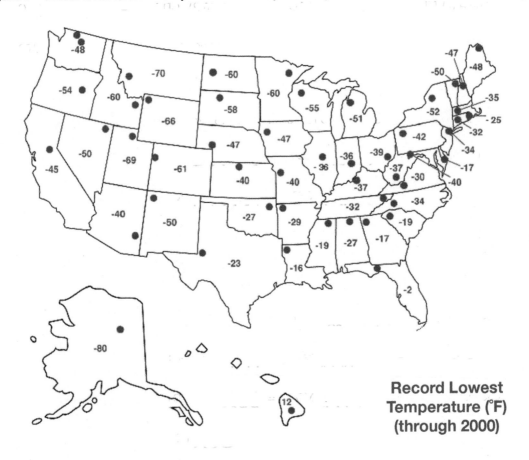

Record Lowest
Temperature (°F)
(through 2000)

11. What are the record temperatures for the state where you were born? (If you were born outside the United States, use the state where you live.)

State _____   Maximum _____   Minimum _____

Part III: Weather Records

12. The largest hailstone ever observed was 18.75 inches in circumference. It fell on 22 June 2003, in Aurora, Nebraska, which is in south-central Nebraska. The previous record was a 17.5-inch hailstone, which fell at Coffeyville, Kansas on 3 September 1970. Coffeyville is in southeast Kansas. Plot 17.5 in southeast Kansas and 18.75 in southern Nebraska on Figure 4.

13. The highest wind speed observed with an anemometer was 231 mph at Mt. Washington, NH on 12 April 1934. Plot 231 in New Hampshire on Figure 4.

14. The highest wind speed observed with a Doppler radar was over Bridge Creek, OK. The radar measured 321 mph on 3 May 1999. Bridge Creek is located about 15 miles southwest of Oklahoma City. Plot 321 at Bridge Creek on Figure 4.

Part IV: Time Conversion

15. Convert the following times to UTC: _____

    a. 2:00 PM PST = _____ UTC   e. 6:00 AM CST = _____ UTC

    b. 12:00 AM PDT = _____ UTC   f. 7:00 PM CDT = _____ UTC

    c. 8:00 PM EST = _____ UTC   g. 12:00 PM MST = _____ UTC

    d. 3:00 PM EDT = _____ UTC   h. 10:00 AM MDT = _____ UTC

16. Convert the following times from UTC to the indicated U.S. time zones:

    a. 1500 UTC = _____ CDT = _____ EDT

    b. 0000 UTC = _____ CST = _____ PDT

    c. 1200 UTC = _____ CDT = _____ MDT

    d. 2100 UTC = _____ CST = _____ EST

    e. 1800 UTC = _____ PDT = _____ EDT

    f. 0715 UTC = _____ EST = _____ CST

    g. 0335 UTC = _____ MDT = _____ CDT

    h. 1101 UTC = _____ EDT = _____ PDT

The Earth-Atmosphere System

LAB ACTIVITY OBJECTIVES:

- You will describe how sunrise and sunset times change with latitude, longitude, and time of year.
- You will discuss how and why the amount of incoming solar radiation is related to latitude and longitude on the earth's surface.
- You will explain the physical reason for the changes in seasons.
- Given a graph of solar radiation, you will determine the times of sunrise and sunset, and you will calculate the length of a day.

MATERIALS NEEDED:

- Laboratory manual
- Pencil or pen
- A flashlight
- Globe

GLOSSARY:

Angle of Incidence
Autumnal Equinox
Insolation
Mesonet
Pyranometer
Radiation

Solar Noon
Solar Radiation
Summer Solstice
Vernal Equinox
Watt
Winter Solstice

BACKGROUND:

The ultimate source of energy for the earth is the sun. All of the earth's weather results from the fact that the sun heats the atmosphere unequally. This unequal heating primarily occurs because (a) the earth is a sphere and (b) the earth is tilted on its axis (at an angle of 23.5°) with respect to the sun (Figure 1). The **angle of incidence** is defined as the angle at which a ray of light (or **radiation**) strikes a surface (Figure 2). It is measured *between* the incoming ray and a perpendicular plane to the surface at the point of incidence (i.e., where the ray strikes). With respect to incoming **solar radiation** striking the earth, the angle of incidence depends on both latitude and time of the year. Hence, the amount of solar energy received at a given location on the earth is related to its latitude and the geometry of the earth's orbit around the sun (Figures 3 and 4).

The time of the year and the latitude of a given location also specify the length of the day. For clear days, weather stations at the same latitude should measure approximately the same amount of total incoming solar radiation and will have the same length-of-day. However, the sunrise and sunset times for these stations will differ according to their longitude.

Same latitude-same Day lenght But Different sunrise/sunset

Same line of longitude-same sunrise/sunset But Different Day lenght

Figure 1 – The Earth's Tilt and Its Orbit Around the Sun

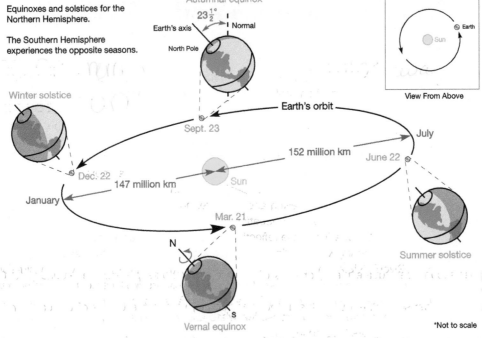

Equinoxes and solstices for the Northern Hemisphere.

The Southern Hemisphere experiences the opposite seasons.

*Not to scale

Figure 2 – The Angle of Incidence (θ)

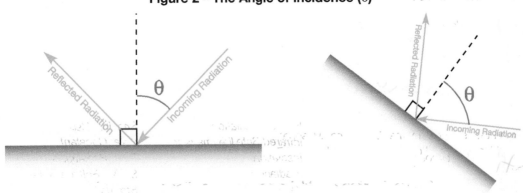

SEASONS

Contrary to most first guesses, seasons result from the tilt of the earth on its axis and not the distance from the sun to the earth (Figure 1). In fact, the sun is *closest* to the earth during the *winter* of the northern hemisphere. Because of the tilt of the earth, the sun appears to migrate between the Tropic of Cancer, at 23.5° N, and the Tropic of Capricorn, at 23.5° S. The **autumnal equinox** and the **vernal equinox** occur when the sun is directly overhead at the equator at **solar noon**. The **summer solstice** for the northern hemisphere occurs when the sun is directly overhead at the Tropic of Cancer (approximately June 21). The **winter solstice** for the northern hemisphere (approximately December 21) occurs when

the sun is directly overhead at the Tropic of Capricorn (i.e., winter occurs in the northern hemisphere because the northern hemisphere is pointed away from the sun). The summer solstice in the *southern hemisphere* occurs during the winter solstice of the northern hemisphere.

During an equinox (Figure 3), less radiation is received at the poles, and more, direct radiation is received at the equator. During a solstice (Figure 4), the summer hemisphere points directly toward the sun and the winter hemisphere has an area (located poleward of the Arctic or Antarctic Circle) that receives no sunlight.

Radiation and Energy Transfer

LAB ACTIVITY OBJECTIVES:

- Given the Stefan-Boltzmann and Wien's Displacement Laws, you will calculate the radiative energy from both the sun and the earth and to calculate the wavelengths of maximum radiation from both bodies.
- Using these same equations and knowledge about the geometry of the sun-earth system, the advanced student will compute the "solar constant" and the irradiance at the top of the earth's atmosphere at different latitudes.
- Using the Stefan-Boltzmann Law and near-surface air temperatures, you will estimate the irradiance at the surface at a specific location.
- Given values of upwelling and downwelling shortwave radiation at the surface, you will compute the albedo at a location.
- Using knowledge of the reflective properties of the earth's surface, you will hypothesize the reason(s) why albedo may change at a location during the year.
- Given visible, infrared, and water vapor satellite images, you will describe where and what type (e.g., cumulus, stratus, cirrus) of clouds are located in a specified region.
- Given a visible satellite image and surface observations (e.g., solar radiation, air temperature), you will describe where and what type (e.g., cumulus, stratus, cirrus) of clouds are located in a specified region.

MATERIALS NEEDED:

- Laboratory manual
- Pencil or pen
- Calculator

GLOSSARY:

Absorption	Infrared Radiation	Sensible Heat
Albedo	Infrared Satellite Imagery	Solar Constant
Blackbody	Insolation	Solar Radiation
Cirrus	Irradiance	Stefan-Boltzmann Law
Condensation	Kelvin	Stratus
Conduction	Latent Heat	Transmission
Convection	Meteogram	Terrestrial Radiation
Cumulus	Power	Troposphere
Downwelling Radiation	Radiation	Upwelling Radiation
Electromagnetic Spectrum	Reflection	Visible Satellite Image
Electromagnetic Waves	Refraction	Water Vapor Image
Evaporation	Satellite Image	Wien's Displacement Law
Heat	Scattering	

BACKGROUND:

Radiation is an essential component of the sun-earth-atmosphere system. Radiation is important because it is one of three processes that transfer energy from one body to another. The other processes are **convection** and **conduction**. Radiation is the only **heat** transfer process mechanism by which the earth receives heat from the sun.

Radiative energy is transferred in the form of **electromagnetic waves** (EM) that move at the speed of light (c; 300 million meters per second). The wavelength (λ) of an electromagnetic wave is measured as the distance from one crest of a radiative wave to the next (Figure 1). The frequency (f) is defined as the rate at which wave crests pass a fixed point. The frequency equals the speed of light divided by the wavelength ($f = c/\lambda$) and has units of s^{-1}, or Hz ("hertz"). For example, suppose your FM radio is tuned to the frequency 99.9 MHz ("megahertz"). Hence, 99.9 million waves from this station pass by your location every second. Using $f = c/\lambda$, the radio station broadcasts its energy with a wavelength of $(99{,}900{,}000\ s^{-1}) / (300{,}000{,}000\ m\ s^{-1}) = 0.33$ m, or about one foot.

Figure 1 – An Electromagnetic Wave

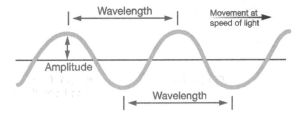

Figure 2 – The Electromagnetic Spectrum

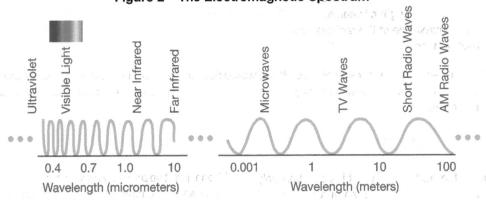

Electromagnetic waves compose a **spectrum** of wavelengths (Figure 2) and include familiar types of radiation: ultraviolet rays, **infrared** (IR) **radiation**, visible light, microwaves, x-rays, TV signals, and radio waves.

Radiation obeys certain laws of physics. Two of these laws are the **Stefan-Boltzmann Law** and **Wien's Displacement Law**. These laws apply to radiative **blackbodies** and they provide us with the following knowledge (e.g., Figure 3):

(a) warmer bodies radiate more energy than do cooler bodies,

(b) warmer bodies radiate at shorter wavelengths than do cooler bodies,

(c) good emitters of energy are good absorbers of radiative energy at a particular wavelength, and

(d) poor emitters of energy are poor absorbers of radiative energy at the same wavelength.

Figure 3 – Schematic of Warm Colors Radiating at Shorter Wavelengths and Higher Energy Than Cool Colors

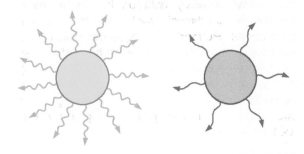

Mathematically, these two laws are described below

Stefan-Boltzmann Law:

where:

$$E^* = \sigma T^4$$

E^* = the blackbody irradiance [in W m⁻²]
σ ("sigma") = 5.67×10^{-8} W m⁻² K⁻⁴ [Stefan-Boltzmann constant]
T = the temperature of the radiating body [in K]

Note: The **Kelvin** temperature scale is an absolute temperature scale in which a change of 1 Kelvin (K) equals a change of 1 degree Celsius. The lowest temperature on the Kelvin scale is 0 K. The melting/freezing point of water is 273 K, or 0°C. The boiling point of water is 373 K, or 100°C.

Wien's Displacement Law:

where:

$$\lambda_{max} = 2897 \text{ [μm K]} / T$$

λ ("lambda") = wavelength [in micrometers (μm)]
λ_{max} = the wavelength of peak emission [in micrometers (μm)]
T = the temperature of the radiating body [in K]
μ ("mu") = micro or one-millionth

Because the sun and the earth generally act like blackbodies, these two laws can be used to calculate the approximate **irradiance** and wavelength of peak emission of both spheres (see questions 1, 3, and 8 in the laboratory exercises).

RADIATIVE PROCESSES

The shape of the earth and its orbital position only determine the *maximum* amount of **solar radiation** that can strike *the top of* the earth's atmosphere at a given latitude and longitude. As the electromagnetic waves travel through the atmosphere, the radiation interacts with gases, dust, clouds, and other atmospheric constituents. Hence, solar radiation measured at the ground often is less than the maximum possible.

Radiation can be absorbed, reflected, refracted, scattered, or transmitted. **Absorption** is the process of retaining energy in a substance. **Reflection** is the process whereby radiation is directed back into the medium through which it traveled (e.g., the atmosphere). **Refraction** changes the direction of electromagnetic (EM) waves as a result of a change in density of the medium or media (e.g., water droplets) through which they travel. **Scattering** is the process by which EM waves are forced to change their direction of motion. Finally, electromagnetic waves are **transmitted** when they pass through space or the media in an unaltered state.

These radiative processes are important to understand because the amount of energy available to heat the earth's surface is related to how much solar radiation is absorbed, reflected, or scattered by the atmosphere before it reaches the surface. In addition, the warmth provided to the earth by its atmosphere (i.e., the "greenhouse effect") depends on the absorption of **terrestrial radiation** (i.e., radiation emitted *from* the earth) by molecules in the atmosphere.

Figure 4 – Reflection, Absorption, Scattering, Refraction, and Transmission of Solar Radiation

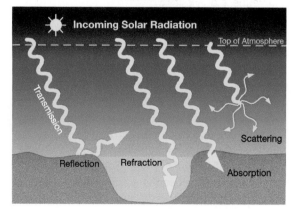

THE ENERGY "BUDGET"

Over long periods of time, the total amount of energy entering the earth's atmosphere must equal the total amount of energy leaving the earth's atmosphere. If the incoming and outgoing energies at the top of the atmosphere were not balanced, then the atmosphere would warm or cool uncontrollably. Atmospheric scientists communicate this energy balance through a global energy "budget," an accounting of the annual mean energy inputs and outputs. Figure 5 displays the portion of the global energy budget relating to the energy *input* to the atmosphere — incoming solar radiation, or **insolation**. Only half of the sun's rays are absorbed at the surface; 30% of insolation escapes back to space; and almost 20% of incoming sunlight is absorbed in the atmosphere.

The energy *output* from the earth and its atmosphere into space (right side of Figure 6) includes solar radiation that is reflected or scattered back to space and terrestrial radiation that is emitted by the earth and its atmosphere. Because of the temperature of the earth and its atmosphere, terrestrial radiation is emitted as infrared radiation.

There is a similar energy balance that is measured over long periods of time at the earth's surface. To keep the surface from heating or cooling uncontrollably, an equal amount of energy must be released from the surface as that absorbed from direct or indirect radiation (see bottom of Figure 5). In this case, most of the outgoing energy from the surface is from infrared radiation (because the earth is warm and radiates energy) and from **latent heat** release (because **evaporation** and **condensation** transfer energy). A small amount of energy also is transferred by conduction between the earth's surface and the air it contacts directly. This energy is called **sensible heat** because you can sense (feel) its warmth (i.e., by placing your hand directly onto a parking lot surface during the heat of summer).

ALBEDO

The **albedo** is the fraction of radiation that reflects off a body. In meteorology, that body typically is the earth's surface or clouds. The values for albedo differ from location to location and even from day to day. The color and texture of the earth's surface affect the albedo at a given location. Plants, clouds, wet soils, and snow influence the daily albedo. The primary result of a greater albedo is a reduced ability for the surface to absorb solar radiation.

Figure 5 – Radiation Budget for Incoming Solar Radiation

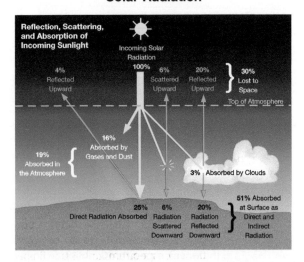

Figure 6 – Globally Averaged Energy Budget

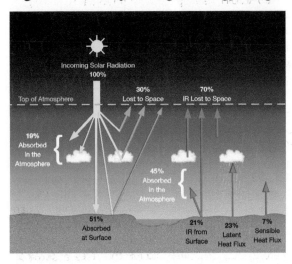

Surface albedo can be calculated from measurements of **downwelling** solar radiation (i.e., the solar radiation aimed *downward toward* the earth's surface) and **upwelling** solar radiation (i.e., the solar radiation reflected *upward from* the earth's surface):

Albedo = (Upwelling radiation) / (Downwelling radiation)

WEATHER SATELLITE IMAGERY

As depicted in Figure 6, radiation interacts with clouds in multiple ways, including reflection of sunlight (e.g., visible light) and emission of infrared waves. In the late 1950s, scientists and engineers at the University of Wisconsin, the National Aeronautic and Space Administration (NASA), and the National Oceanic and Atmospheric Administration (NOAA) used this basic physics knowledge to initiate one of the greatest technological advances in meteorology: the weather satellite. Today, satellites gather most of the world's weather data.

The three types of **satellite images** most used by meteorologists are visible, infrared, and water vapor images. Instruments on the satellite sense different wavelengths in the EM spectrum to obtain these images. Visible satellite images capture details in the 0.3 to 0.7 µm band, covering the visible portion of the spectrum (see Figure 2). Infrared satellite images sense radiation between 10.2 and 11.2 µm. Water vapor images also measure radiation emitted in the infrared, but focus on the 6 to 7 µm band, where water vapor in the *upper half* of the **troposphere** can be sensed.

Visible satellite images essentially capture a grayscale photograph of what you would see if you took a snapshot from the satellite's altitude (35,800 km, or 22,300 miles, above the surface). White and light gray areas are highly reflective surfaces, such as clouds or snow; dark gray areas usually indicate land surface or oceans. Visible imagery can reveal areas of differing surface characteristics (e.g., water bodies, deserts) and differing types of cloud features (e.g., **cumulus** clouds that are associated with low-level moisture, low-level **stratus** clouds and fog, or high-level **cirrus** clouds).

Infrared satellite imagery displays radiation from the relatively wide temperature range we could sense throughout a transcontinental airplane flight, including warm temperatures near the surface and very cold temperatures at the top of the troposphere. On an infrared, or IR, image, dark gray areas are warm and typically represent the earth's surface or low clouds. Light white areas are cold and typically depict cirrus clouds or the upper portions of tall cumulonimbus clouds.

Water vapor images have multiple uses. First, these images directly provide information on the water vapor content of the *upper half* of the troposphere. Water vapor imagery gives little information about moisture near the surface. The darkest areas on water vapor imagery reveal areas that are very dry throughout the upper atmosphere. Lighter areas highlight locations where upper-level moisture is abundant. Second, the dark (dry) areas often mark sinking motion in the atmosphere, where air compresses and, as a result, warms. Third, because water vapor can act as a tracer of atmospheric motion, water vapor images sometimes can provide details of atmospheric circulation patterns that cannot be observed by any other means.

LABORATORY EXERCISES:

Part I: Radiation from the Sun

1. The average temperature of the sun is 5780 K. Using the Stefan-Boltzmann Law, calculate the average *irradiance* of the sun. Show your work and include appropriate units.

$$E = 6.3 \times 10^7 \, W/m^2$$

~~$W = 4 \times 10^{26}$~~

2. The radius of the sun is 7×10^8 meters. How much total **power** is emitted from the sun? (Hint: The surface area of a sphere is equal to $4\pi r^2$, where r is its radius.) Show your work and include appropriate units.

$$W = 4 \times 10^{26}$$

3. Using Wien's Displacement Law and information from Question 1, (a) calculate the wavelength of peak emission of sunlight. Show your work and include appropriate units. Using Figure 2, (b) list the primary type of radiation that the sun emits (e.g., ultraviolet, visible, infrared, etc.).

$$.05 \mu m = visible$$

Primary Type of Radiation __visible__

4. **(Advanced Students/Meteorology Majors)** The **solar constant** is the power per unit area (irradiance) striking an imaginary flat plane that is perpendicular to the incoming solar rays at the top of the earth's atmosphere. Mathematically, the solar constant is denoted by S_0. The solar constant can be calculated using the following inverse square law:

$$S_0 = E_{sun} \times (R_{sun}/r)^2$$

where:

E_{sun} = the irradiance from the sun [in W m^{-2}],
R_{sun} = the radius of the sun = 7×10^5 km, and
r = the mean distance between the earth and the sun = 1.5×10^8 km.

Calculate S_0. Show your work and include appropriate units.

Figure 10 – Infrared Satellite Imagery for 11 March 2000 at 1745 UTC (11:45 AM CST)

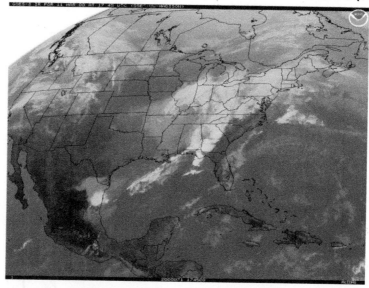

Figure 11 – Central Plains Visible Imagery for 11 March 2000 at 1732 UTC (11:32 AM CST)

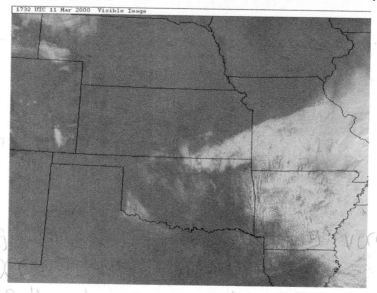

Figure 12 – Incoming Solar Radiation (in W m⁻²) for 11 March 2000 at 11:30 AM CST (1730 UTC), as Measured by the Oklahoma Mesonet

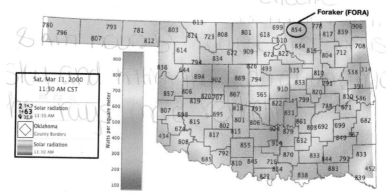

Figure 13 – Meteogram from Foraker (OK) Mesonet Station for 10-11 March 2000

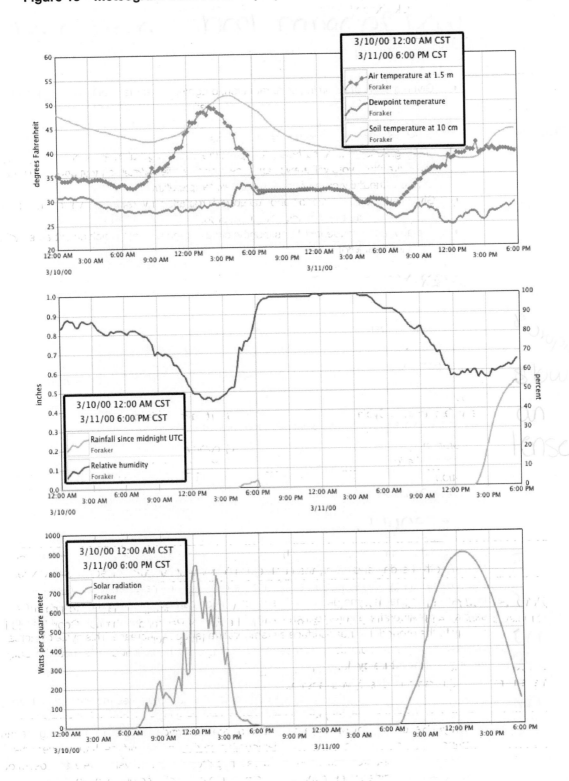

Daily Temperature Cycle

LAB ACTIVITY OBJECTIVES:

- Given graphs of incoming solar radiation, outgoing terrestrial radiation, and near-surface air temperature, you will describe how the daily temperature cycle relates to radiation.
- You will describe how secondary factors, such as wind, soil type, and atmospheric moisture, affect the daily temperature cycle.
- Given graphs of near-surface air temperature, incoming solar radiation, wind speed, and wind direction, you will analyze and describe the effects the radiation and wind variables have on the daily cycle of near-surface air temperature.
- Given sufficient information about a specific radiation inversion, you will deduce how the inversion affects the farming of a given crop.
- The advanced student will describe the circumstances under which an extreme radiation inversion will form.

MATERIALS NEEDED:

- Laboratory manual
- Pencil or pen

GLOSSARY:

Albedo
Daily Temperature Cycle
Diurnal
Evaporation
Evapotranspiration
Insolation
Longwave Radiation

Radiation Inversion
Shortwave Radiation
Solar Radiation
Specific Heat
Terrestrial Radiation
Vertical Temperature Profile

BACKGROUND:

During the day, the sun's rays warm the ground, leading to an increase in energy at the ground. During both the day and night, the earth emits **terrestrial radiation**, leading to a decrease in the surface energy. Because the daytime increase in surface energy caused by **solar radiation** is *greater than* the daytime decrease in surface energy caused by terrestrial radiation, the near-surface air temperature tends to increase during the day. At night, when the sun does not add energy to the ground, the near-surface air temperature tends to decrease. This cycle of increasing and decreasing near-surface air temperatures during a 24-hour period is termed the **daily** (or **diurnal**) **temperature cycle**.

As you know from previous lessons, the primary factor that influences the amount of incoming solar radiation (i.e., **insolation**) and the daily temperature cycle is the angle of incidence of the sun's rays. Hence, the day of the year and latitude are two critical factors in determining the range of temperatures experienced at a particular location on the earth's surface. If these were the only factors, then temperature forecasting would be straightforward. However, there are a host of other factors that cause differences in the daily temperature cycle at different locations.

Variables such as soil type, soil moisture, vegetation, elevation, and proximity to water contribute to the daily variation of temperature. In addition, the wind causes changes in humidity, cloud cover, and rainfall that affect air temperatures significantly. Several of the most important influences are highlighted below:

Soil Type – The **albedo** of bare soil depends on the soil type — its color, in particular. White sand reflects most of the incoming solar radiation (i.e., high albedo) whereas black loam absorbs most insolation. Hence, in the absence of any other differences, the range of temperatures will be greater over black loam than over white sand. The albedo of bare soil varies between these two extremes and is an important factor to forecast temperature accurately over non-vegetated areas during the summer.

Soil Moisture – Water has a high heat capacity, or **specific heat**, as compare to air. In other words, given the same amount of energy added to water and to air, water heats significantly more slowly than air. Hence, if water occupies the space between soil particles, then the soil warms more slowly than if air occupies that same space (i.e., the soil is dry). In addition, at the ground surface, a moist soil evaporates water into the air. Because **evaporation** is a cooling process, the near-surface air temperature will remain cooler during the day over a moist soil than over a dry soil. Evaporation is a cooling process; it requires the air to supply latent heat to initiate the phase change between liquid and vapor. The effect of the evaporation of moisture from the soil into the air is that the near-surface air temperature remains cooler during the day over a moist soil than over a dry soil.

Vegetation – Vegetation has several influences on the daily temperature cycle. First, vegetation may change the surface albedo and, thus, the absorption of solar energy. In general, vegetation has a low albedo (i.e., high absorption of insolation) and can change the albedo of an area profoundly if vegetation covers a soil type with a moderate or high albedo. Leafy surfaces hold water after rain events or dew formation until the water evaporates, enabling the cooling effect of evaporation. Additional cooling is provided by the plant's transpiration — a process closely linked to the availability of soil moisture in the root zone. The combined effect of evaporation and transpiration is called **evapotranspiration**.

Clouds – Clouds affect daily temperatures in multiple ways. Figures 5 and 6 from the lab entitled "Radiation and Energy Transfer" depict many of the interactions between clouds and radiation. For example, clouds reflect incoming solar radiation to space. Thicker clouds reflect more sunlight, leading to cooler surface temperatures during the day. Clouds also absorb upwelling longwave radiation from the earth and emit some of the energy back to the ground. This insulating effect of clouds is most obvious during the night, when temperatures remain warmer than if the skies were clear. Low, thick clouds (e.g., fog, stratus) emit more radiation toward the ground, leading to warmer surface temperatures during the night. Clouds also can bring precipitation that moistens the soil and, in some cases, changes the albedo of the surface (e.g., snow).

Humidity – The daily temperature range is less in a humid location than in a dry one. Water vapor in the air absorbs both solar and terrestrial radiation and emits longwave radiation back toward the ground. Therefore, daytime temperatures at humid locations typically are cooler than sites with low humidity; nighttime temperatures in humid regions remain warmer than drier areas.

VERTICAL TEMPERATURE PROFILES NEAR THE SURFACE

The earth and the air do not absorb radiation at the same rate, leading to discontinuities between the temperatures of the ground and the air close to the ground. These discontinuities influence how we live at the interface between the atmosphere and the earth's surface.

During a sunny day, solar radiation heats the ground and, in turn, the ground heats the air within the closest few centimeters. Like rising bubbles in a boiling pot of water, this warmer air near the ground can rise to transport heat to layers well above the surface. These *convective thermals* increase in height as the ground continues to warm through early or mid afternoon. Like stirring the boiling water in the pot, the wind mixes the lower atmosphere and distributes temperature more evenly, reducing the differences in temperature with height. Figure 1 shows a typical **vertical temperature profile** during both a calm day and a windy day.

During a clear night, the ground radiates energy faster than the air above. Thus, air near the ground cools significantly faster than the air above, leading to a **radiation inversion**. Figure 2 demonstrates an example of a temperature inversion at night.

10. Look at the plots below of air temperature, solar radiation, and winds measured by ARM at Barrow, AK (Figures 6-8). For each of the three plots, document any time lag between observed changes in solar radiation and corresponding changes in air temperature. Summarize your observations by describing the magnitude of the lag as a function of season, cloudiness, and speed of the winds. If the time lag in any of the plots does not agree with your expectations, provide a hypothesis as to why the expected relationship did not occur.

_____

_____

_____

_____

_____

_____

Figure 6 – Air Temperature (in °C), Solar Radiation (in Wm⁻²), Wind Speed (in m/s), and Wind Direction (in deg from North) from the ARM Climate Research Facility at Barrow, AK on 26 April 2002

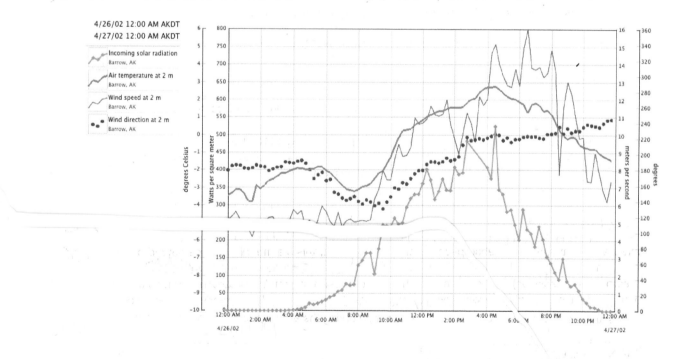

Figure 7 – Air Temperature (in °C), Solar Radiation (in Wm⁻²), Wind Speed (in m/s), and Wind Direction (in deg from North) from the ARM Climate Research Facility at Barrow, AK on 20 June 2003

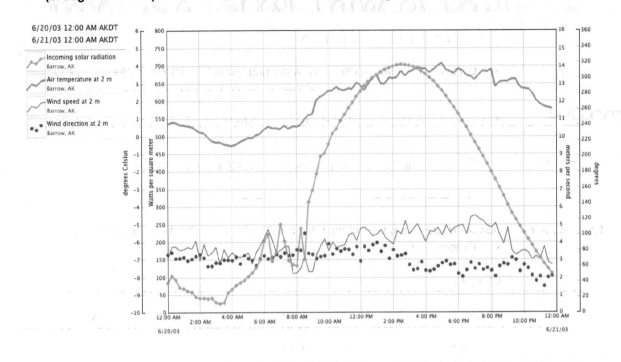

Figure 8 – Air Temperature (in °C), Solar Radiation (in Wm⁻²), Wind Speed (in m/s), and Wind Direction (in deg from North) from the ARM Climate Research Facility at Barrow, AK on 3 October 2003

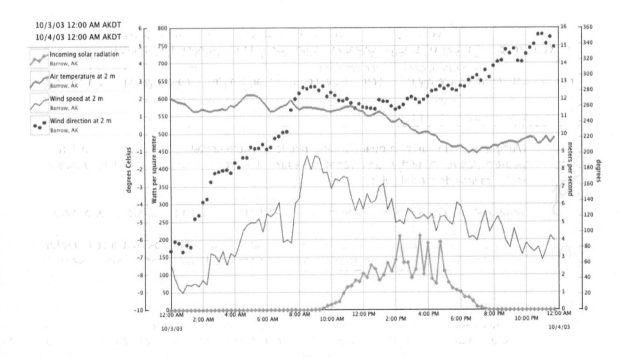

Atmospheric Moisture

LAB ACTIVITY OBJECTIVES:

- Given air temperature, dew point, and a graph of saturation vapor pressure as a function of temperature, you will determine the saturation vapor pressure, vapor pressure, and relative humidity.
- Given different scenarios associated with evaporation and condensation, you will explain why evaporation or condensation occurred.
- Using a set of dewpoint temperature and wind vector maps, you will determine regions of moist air advection, dry air advection, and no moisture advection.

MATERIALS NEEDED:

- Laboratory manual
- Pencil or pen

GLOSSARY:

Advection	Latent Heat	Temperature
Air Parcel	Prevailing Wind Direction	Vapor Pressure
Condensation	Relative Humidity	Vector
Cumulonimbus Cloud	Saturation (of Air)	Virga
Dalton's Law of Partial Pressures	Saturation Vapor Pressure	Wind Direction
Dew Point (or Dewpoint Temperature)	Sensible Heat	Wind Speed
Evaporation	Specific Heat	

BACKGROUND:

Water vapor in the atmosphere is important, not only because it is the raw material for precipitation, but also because it carries large amounts of **latent heat** (i.e., energy) and is an efficient greenhouse gas. These properties allow water vapor to play a major role in many meteorological processes, even though water molecules are only a small fraction of the atmosphere (1-4%). It is for these reasons that it is important to learn about moisture and how it is measured.

Multiple variables exist to quantify atmospheric moisture content. These variables include **dew point**, **relative humidity**, and **vapor pressure**. Important concepts related to atmospheric moisture include the following:

Temperature (T) – The air temperature determines the maximum amount of water vapor molecules that can exist in the atmosphere. More water vapor *can be present* in warm air than in cool air. However, just because warmer air allows for more water vapor, hot air does not necessarily have a high water vapor content. Air temperature simply determines the *maximum amount of water vapor possible*, not the actual amount.

Dew Point, or Dewpoint Temperature (T_D) – The dew point is one measure of the actual amount of water *vapor* in the air. It is defined as the temperature to which the air must be cooled for **saturation** to occur. When saturation occurs, no additional water vapor molecules can be added to the air mixture; hence, any additional water molecules will condense to form *liquid* droplets. Higher dew points mean more moisture is present in the air. Because the air temperature determines the maximum amount of water vapor possible, the dewpoint temperature typically is not greater than the air temperature (except for rare "supersaturated" conditions). At saturation, the air temperature and the dewpoint temperature are identical values. Also, if the air is saturated and cooling occurs, then some of the water vapor molecules are forced to condense into liquid water. Dew point typically is expressed in degrees Celsius or Fahrenheit.

Vapor Pressure (e) – Vapor pressure is the contribution to the total atmospheric pressure (in **millibars**, or mb) by water vapor molecules. Because air is a mixture of gases, each constituent gas, including water vapor, exerts its own partial pressure. The sum of the partial pressures of all gases is the total air pressure. This concept is known as **Dalton's Law of Partial Pressures**. [For example, assume the total atmospheric pressure is 975 mb. This pressure could result from partial pressures of 761 mb from nitrogen, 205 mb from oxygen, 6 mb from water vapor, and 3 mb from carbon dioxide — totaling 975 mb. In this case, the vapor pressure (e) is equal to 6 mb.] Larger values of vapor pressure reflect higher moisture content of the atmosphere. Vapor pressure is solely a function of dewpoint temperature, depicted mathematically as $e(T_D)$. Vapor pressure typically is expressed in millibars (mb).

Saturation Vapor Pressure (e_s) – The **saturation vapor pressure** is the vapor pressure of the air at saturation. It represents the maximum vapor pressure that the air could have at the given temperature. Saturation vapor pressure is solely a function of air temperature, depicted mathematically as $e_s(T)$. Saturation vapor pressure typically is expressed in millibars (mb).

Figure 1 displays how the saturation vapor pressure changes as a function of temperature. For example, if the air temperature were 30°C, then e_s = 42 mb.

Relative Humidity (RH) – Relative humidity (RH) is the ratio of the actual vapor pressure (e) to the saturation vapor pressure (e_s). Mathematically, RH = (e/e_s)x100%. Hence, relative humidity quantifies how close the air is to being saturated. If the relative humidity is 100%, then the air is saturated. If the RH is 0%, absolutely no water vapor is in the air (not realistic). If the RH is 50%, then the air contains 50% of the water vapor that it could at a given temperature. It is important to note that the atmosphere is always unsaturated unless its RH is 100%. That means very warm air with a RH of 99.9% is considered unsaturated. Relative humidity depends on both the air temperature and its dew point. Thus, a change in either variable will change the resulting relative humidity.

Figure 1 – Saturation Vapor Pressure as a Function of Temperature

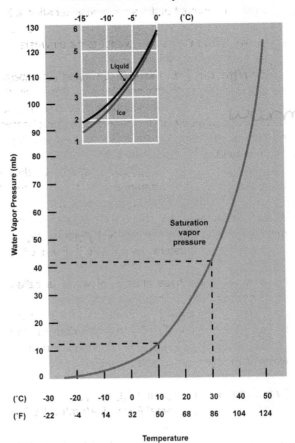

Figure 1 can be used to calculate the relative humidity given both the air temperature and the dew point. Suppose that you measure an air temperature of 30°C and a dew point of 10°C. Use Figure 1 to find the value of e_s that corresponds to your measured value of air temperature. Also use Figure 1 to find the value for e that corresponds to your measured value of dew point. Hence, e_s = 42 mb and e = 12 mb. To compute RH, divide e by e_s (12 mb / 42 mb = 0.29) and multiply by 100%. For this combination of air temperature and dew point, the relative humidity is 29%.

The National Weather Service and other weather networks use a similar process daily. Figure 1 indicates that you can calculate any one of the variables T, T_D, or RH if you know the other two variables. Some weather networks use instruments to measure T and T_D, and then calculate RH. Other networks have instruments that measure T and RH, and then calculate the dew point.

Air Masses and Fronts

LAB ACTIVITY OBJECTIVES:

- You will describe the important characteristics of air mass types.
- Using a map of the continental United States, you will label the source regions of the air masses that influence weather in the U.S.
- Given a set of surface weather maps that depict a front, you will identify and name two distinct air masses.
- Given a sequence of surface weather maps that include several atmospheric variables (e.g., temperature, rainfall) associated with a frontal passage, you will locate the front, determine the type of front, determine its direction of movement, and list at least five weather conditions (e.g., cloudy behind the front) associated with the specific front.

MATERIALS NEEDED:

- Laboratory manual
- Pencil or pen
- Colored pencils

GLOSSARY:

Air Mass
Cold Front
Continental Arctic Air Mass
Continental Polar Air Mass
Continental Tropical Air Mass

Front
Maritime Polar Air Mass
Maritime Tropical Air Mass
Warm Front

BACKGROUND:

An **air mass** is a large body of air with similar properties of temperature and humidity. Air masses are named after the source region where they originate.

When a large body of air (e.g., >500 km across) remains over an area for several days or weeks, the body of air often acquires the thermal and moisture properties of the underlying land or water. The air exchanges heat with the land or water, either warming or cooling, until its temperature closely matches that of the surface below it. The body of air also gains or loses moisture depending on the air temperature and moisture content of the surface. Table 1 lists characteristics of air mass source regions.

Table 1 – Characteristics of Air Mass Source Regions

Source Region	Air Mass Characteristic	Symbol
Continental (land)	Dry	c
Maritime	Moist	m
Polar	Cool/Cold	P
Arctic	Extreme Cold	A
Tropical	Warm	T

Air mass names are constructed by combining the temperature and moisture characteristics, resulting in five air mass types (Table 2). Inherently, each type of air mass has different density properties because *warm air is less dense than cold air and moist air is less dense than dry air.*

FRONTS

A **front** is the transition zone between two different air masses, each having different densities. Thus, fronts usually separate air masses of different temperatures and moisture characteristics (e.g., cold/dry versus warm/moist). On each side of the front, weather patterns reflect the characteristics of the particular air mass on that side. When weather fronts move, one air mass is advancing and the other is retreating.

Four types of fronts are used in operational meteorology: cold front, warm front, stationary front, and occluded front (Figure 1). Occluded and stationary fronts are depicted in Figure 1 but are not discussed in this laboratory exercise. Fronts are identified by their properties (i.e., the air masses they separate and their motion).

Much of the early research regarding fronts was accomplished in Europe during World War I (WWI). Because a front is a "battle zone," where one air mass attempts to overtake another air mass, it was natural for the WWI-era scientists to use the term "front" to describe this zone.

Figure 1 – Map Symbols Used to Depict Different Types of Fronts

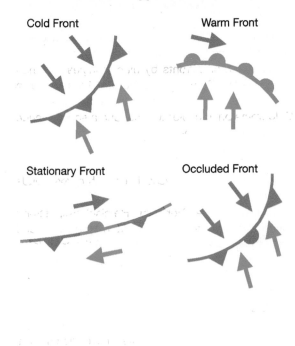

Cold Front   Warm Front

Stationary Front   Occluded Front

Table 2 – Air Mass Classification

Temperature / Moisture	Polar or Arctic	Tropical
Continental	**Continental polar** (cP) *cold and dry* or **Continental arctic** (cA) *extremely cold and dry*	**Continental tropical** (cT) *warm and dry*
Maritime	**Maritime polar** (mP) *cool and moist*	**Maritime tropical** (mT) *warm and moist*

Figure 6 – Dew Point (in °F) for 1 November 2004 at (a) 3:00 AM CST, (b) 9:00 AM CST, (c) 3:00 PM CST, and (d) 9:00 PM CST, as Measured by the NWS

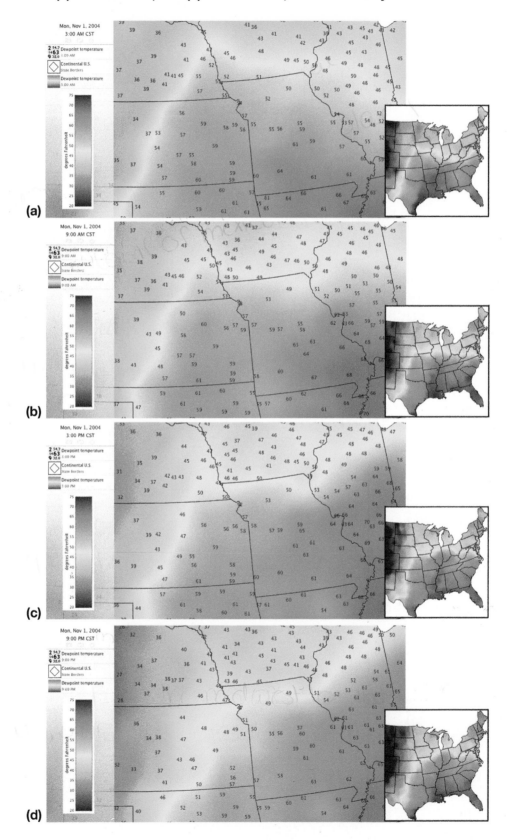

(a)

(b)

(c)

(d)

Figure 7 – 3–Hour Rainfall (in inches) and Wind Vectors for 1 November 2004 at (a) 3:00 AM CST, (b) 9:00 AM CST, (c) 3:00 PM CST, and (d) 9:00 PM CST, as Measured by the NWS

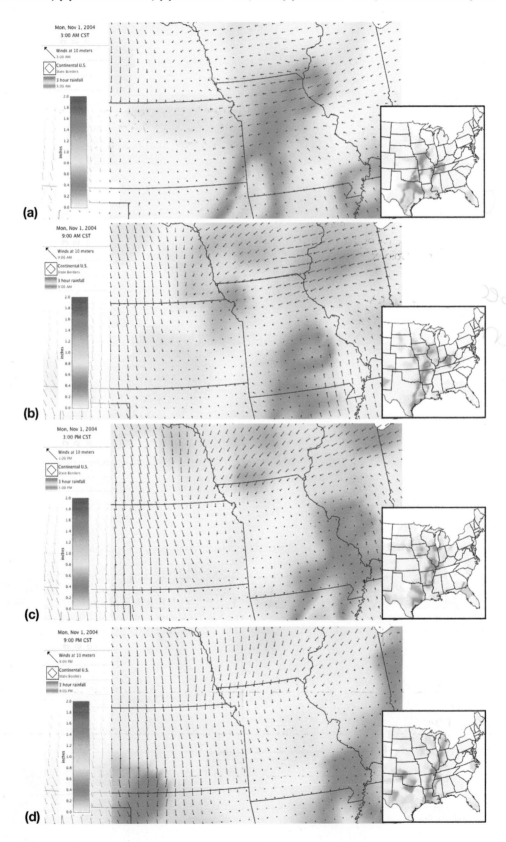

Surface Map Analysis

LAB ACTIVITY OBJECTIVES:

- Given a surface map plotted with the station model format, you will interpret the data and create contour maps applying the data.
- Using a map that you contoured, you will find the strongest gradient in the contoured field.
- The advanced student will decode and create METAR reports.
- Using METAR reports, the advanced student will plot data using the station model.

MATERIALS NEEDED:

- Laboratory manual
- Pencil or pen
- Colored pencils

GLOSSARY:

Altimeter Setting	Mesoscale
Cold Front	Scalar Field
Contour Line	Station Model
Gradient	Surface Map
Isobar	Synoptic Map
Isodrosotherm	Synoptic Scale
Isopleth	Thermal Ridge
Isotach	Warm Front
Isotherm	Wind Barb

BACKGROUND:

Every hour of every day, surface observing stations worldwide measure meteorological variables such as temperature, dew point, wind speed and direction, and pressure. The U.S. National Weather Service (NWS) automatically acquires surface observations using the Automated Surface Observing System, or ASOS. ASOS consists of more than 850 automated stations that are located primarily at airports and distribute their observations at least once per hour. Some ASOS reports are augmented with human observations. ASOS and other surface observing systems around the world create large volumes of data. As a result of the large data volume and poor telecommunications in some countries, the World Meteorological Organization (WMO) developed a universal method for encoding these data in a compressed, teletype-based format (e.g., all uppercase, limited punctuation). This format, called METAR, was adopted by the United States on 1 July 1996, to share data both domestically and among the nations.

THE STATION MODEL PLOT

Hourly **surface maps** are plotted from the decoded METAR observations. These hourly maps are sometimes called **synoptic maps** — maps representing a single time. Meteorologists plot surface weather data on maps in a consistent manner called a **station model**, which has been used for decades. Figures 1–5 illustrate how surface observations are depicted graphically on a station model plot. Figure 1 not only shows the variables plotted, but also the commonly accepted location where these variables appear on the station model. The graphical display of wind speed and direction is called a **wind barb**. As Figure 5 indicates, wind barbs display the wind speed in an additive fashion; hence, 75 knots is represented by a combination of one flag (50 kt), two long barbs (10 kt each), and one short barb (5 kt). In addition, the wind barbs point in the direction from which the wind blows. Some people mentally visualize an arrowhead on the opposite end of the barbs (at the station circle) to interpret the wind direction.

After decoding the data, a station model is plotted on a map at the location of all reporting stations. Figure 6 displays the 3-letter abbreviations for NWS surface observing stations and Figure 7 shows a sample plot of station models for a given date and time.

Figure 1 – The Station Model

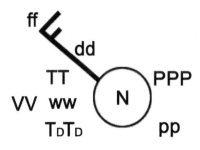

TT – Air temperature (in °F) [Note: °F = (9/5) °C + 32°F]; typically drawn in red

T_DT_D – Dewpoint temperature (in °F); typically drawn in green

ww – Current weather conditions (see Figure 2); typically drawn in black

VV – Visibility (in statute miles); typically drawn in black

dd – Wind direction (the line represents the direction from which the wind is blowing; see Figure 4); typically drawn in black

ff – Wind speed in knots (half line = 5 knots, full line = 10 knots, flag = 50 knots); typically drawn in black

PPP – Sea-level pressure with the leading 9 or 10 removed (in tenths of a millibar); typically drawn in black or blue

pp – Pressure tendency (change in sea-level pressure during the past 3 hours, in tenths of a millibar); typically drawn in black

N – Cloud cover (see Figure 5); typically drawn in black

Figure 2 – Precipitation Symbols Used on a Station Model

Phenomenon	Symbol	Intensity		
		Light	Moderate	Heavy
Rain	• •	• •	•·•	··•·
Snow	* *	* *	*·*	*·*
Drizzle	, ,	, ,	,',	,',,
Rain Shower	▽̇	▽̇	▽̇	
Snow Shower	▽̇	▽̇	▽̇	
Freezing Rain	∿	∿	∿	
Freezing Drizzle	∿	∿	∿	

Figure 3 – Other Weather Symbols Used on Weather Maps

Other Weather Symbols			
Phenomenon	Symbol	Phenomenon	Symbol
Thunderstorm	R	Thunderstorm with Hail	△R
Severe Thunderstorm	ß	Hurricane	∮
Dust Storm	S→	Drifting or Blowing Snow	─+─
Smoke	⌐∿	Haze	∞
Sleet	△	Fog	≡
Tornado or Funnel Cloud	)(	Thunderstorm with Snow	*R
Cold Front (Surface)	▲▲▲	Warm Front (Surface)	●●●
Occluded Front (Surface)	▲●▲	Stationary Front (Surface)	▲●▲

Figure 4 – The Display of Wind Speed on the Wind Barb of a Station Model

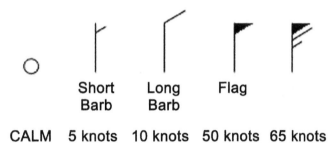

CALM   5 knots   10 knots   50 knots   65 knots

Short Barb   Long Barb   Flag

Figure 5 – The Display of Sky Cover on a Station Model

Amount of Sky Covered

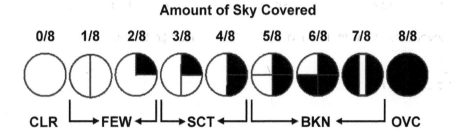

0/8   1/8   2/8   3/8   4/8   5/8   6/8   7/8   8/8

CLR → FEW ←→ SCT ←→ BKN ←─ OVC

Figure 6 – Map of 3-Letter IDs for NWS Surface Stations

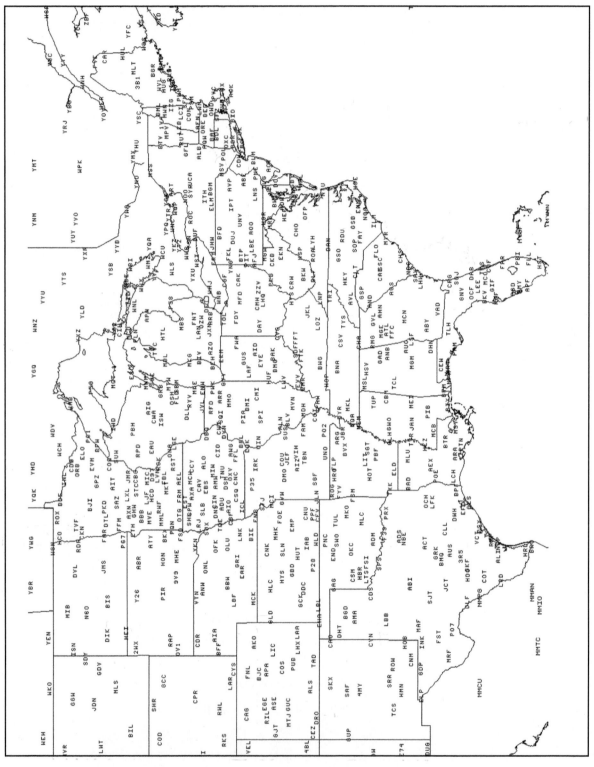

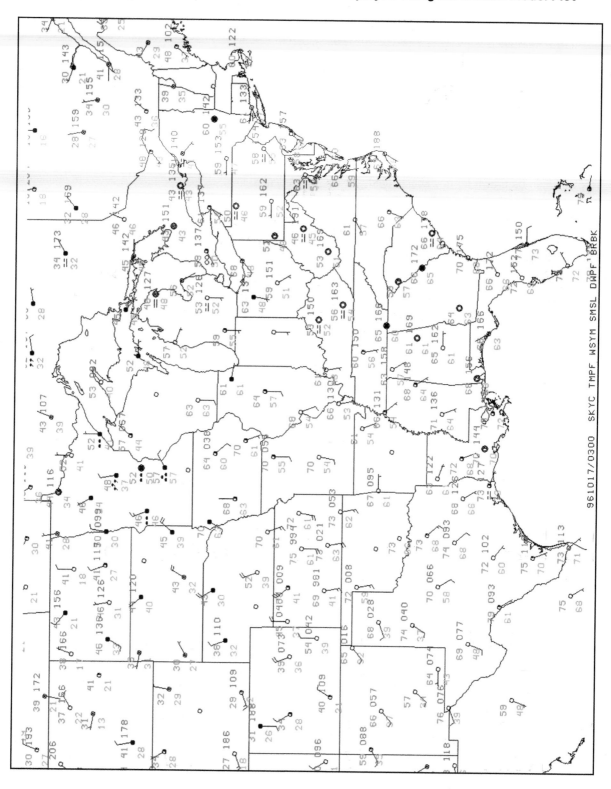

CREATING A CONTOUR MAP

After plotting the data on a map using the station model, meteorologists analyze the observations (e.g., temperature, dewpoint, or pressure) by drawing lines of constant value, known as **isopleths** ("iso" means equal and "pleth" means value) or **countour lines**. Isopleths connect points of equal value and they are drawn at regular intervals. Isopleths separate larger values of a field from smaller values to help the meteorologist visualize the patterns in a given field.

Isopleths can be drawn for any **scalar field**. Typical meteorological variables and their associated isopleths include the following:

Meteorological Variable	Isopleth Name
Temperature	**Isotherm**
Dew Point	**Isodrosotherm**
Pressure	**Isobar**
Pressure Tendency (Change)	Isallobar
Wind Speed	**Isotach**
Wind Direction	Isogon
Rainfall	Isohyet

The process of contouring a scalar field on a weather map aids in the visual interpretation of the data because the process reveals spatial patterns. For example, a contoured field of air temperature can indicate the location of features such as a **cold front**, **warm front**, or **thermal ridge**.

How to Contour a Field of Weather Data

To create the best possible contour analysis, work carefully, use a consistent set of contour intervals, and draw upon your physical understanding of the situation, including the geography surrounding the measurement station (e.g., terrain, land use) and the meteorology behind the event. It is often easiest to begin an analysis in the center of the map rather than at the edge. Be careful to interpolate between observation points in the placement of a contour line to produce a smooth representation of the pattern.

The best contour interval for a specific analysis depends on both the field to be analyzed and the spatial scale of the analysis. For example, a **synoptic-scale** analysis of surface air temperature across the continental U.S. could be performed using a 5°F contour interval without using either too many or too few contour lines. On the other hand, a **mesoscale** analysis of temperatures across a small region, like a state, may use a contour interval of 1 or 2°F so that there are enough isopleths to distinguish a pattern.

Choose contour values that are evenly divisible by the selected interval. For example, a series of isotherms for a temperature analysis using a 5°F contour interval would include 35°F, 40°F, 45°F, and 50°F, rather than 32°F, 37°F, 42°F, 47°F, and 52°F. Typical contour intervals are shown below:

Meteorological Variable	Contour Interval	Suggested Contour Values
Temperature	5°F	50, 55, 60, 65, 70, 75, 80
Dew Point	5°F	40, 45, 50, 55, 60, 65, 70
Sea-Level Pressure	4 mb	992, 996, 1000, 1004, 1008, 1012

The human's understanding of meteorological phenomena is important to accurately interpret the physical situation on a single map and through time from one map to the next. Computer analyses follow a set of basic rules, providing a good "first guess" to the analysis. Computers, however, do not interpret how related meteorological fields may affect the analysis, nor do computers verify that a series of successive maps are internally consistent. Hence, the human can interpret the affect of thunderstorms on the contours of temperature, dew point, pressure, rainfall, and wind. Humans also can verify that temperature **gradients** (e.g., along fronts) do not disappear and reappear through a series of successive maps. The meteorologist's knowledge of the current and recent weather is critical to obtaining the best map analysis. The best forecasters hand-analyze their own maps to supplement the computer-generated analyses.

Guidelines for Contouring Weather Data

The following guidelines will help you draw correct and consistent isopleths of any meteorological field:

1. Begin contouring by lightly sketching the isopleths; then go back and darken them when you are sure they are correct.

2. Isopleths of different values should <u>never</u> cross or branch (i.e., fork). They either should be closed lines or they should end at the edge of the data field.

3. Do not draw an isopleth where no data are available. For example, where weather information plotted on a map is absent over the ocean, the isopleth should end at the last observation over land.

4. Draw isopleths smoothly unless additional information suggests otherwise. For example, radar or satellite imagery can help provide the locations of thunderstorms, fronts, the centers of high- and low-pressure systems, and other features that may affect the analysis.

5. Draw isopleths at equal intervals. Typically, isobars (lines of constant pressure) are drawn at intervals of 4 mb and isotherms (lines of constant temperature) are drawn every 5°F.

6. Label isopleths clearly. Each contour line should be labeled at the beginning and end of the line. Long contour lines also may be labeled at several places along the contour. Closed isopleths can be labeled at any location along the isopleth, but generally the labels appear close to the locations of labels on nearby open isopleths.

7. Locate the maximum and minimum values by examining closed contours. On pressure charts, the maximum value should be labeled with an uppercase "H" (for "high") and the minimum value with an uppercase "L" (for "low"). Maxima and minima in temperature fields are sometimes marked with "W" (for "warm") and "C" (for "cold"). Moisture plots may be labeled with "M" (for "moist") and "D" (for "dry").

LABORATORY EXERCISES:

Part I: Analyses of Station Model Plots

1. You will see both 3-letter and 4-letter station identifications (IDs) on weather maps and tables of data. Four-letter IDs of stations in the continental U.S. begin with a "K," followed by the 3-letter IDs often seen elsewhere. The 3-letter ID typically is an abbreviation for the name of a city or airport where the reporting station is located. For example, KTLH is the ID for Tallahassee, FL. Match the following station IDs with its location (i.e., city and state).

1.	KPHX	_____	a. Atlanta, GA (Hartsfield – Jackson International Airport)
2.	KDFW	_____	b. Miami, FL International Airport
3.	KMIA	_____	c. Tinker Air Force Base, OK
4.	KPHL	_____	d. Minneapolis – St. Paul, MN International Airport
5.	KMSP	_____	e. San Francisco, CA International Airport
6.	KATL	_____	f. Chicago, IL (O'Hare International Airport)
7.	KJFK	_____	g. Dallas – Fort Worth, TX International Airport
8.	KSFO	_____	h. New York City, NY (John F. Kennedy International Airport)
9.	KORD	_____	i. Phoenix, AZ (Phoenix Sky Harbor International Airport)
10.	KTIK	_____	j. Philadelphia, PA International Airport
11.	KSLC	_____	k. Baltimore, MD – Washington, DC International Airport
12.	KBWI	_____	l. Salt Lake City, UT International Airport

2. Using Figure 8, draw isotherms at intervals of 5 degrees Fahrenheit. Be sure to label both ends of the isopleth. The numbers represent air temperature in degrees Fahrenheit. The dots represent the locations of the observing sites.

Figure 8 – Student Exercise for Drawing Isopleths

•	•	•	•	•	•
31	40	46	43	37	41
•	•	•	•	•	•
32	47	53	49	39	42
•	•	•	•	•	•
48	52	61	55	43	51
•	•	•	•	•	•
56	61	68	63	52	55

3. Use the guidelines for contouring weather data to draw isobars at intervals of 4 mb on Figure 9.

Figure 10 – Map of Decoded NWS Surface Data for 17 October 1996 at 0300 UTC for Air Temperature Analysis

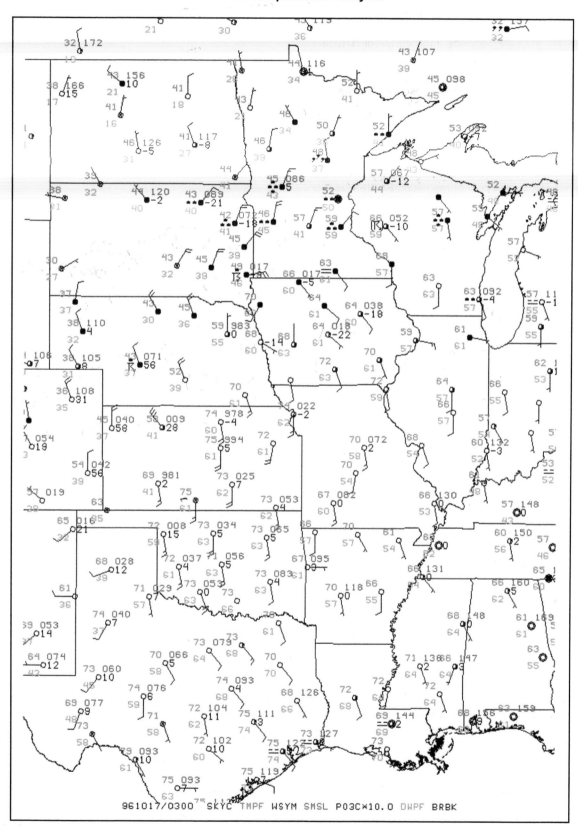

961017/0300 SKYC TMPF WSYM SMSL P03C×10.0 DWPF BRBK

Part II: Decoding METARs

This section is recommended for advanced students and meteorology majors only. Decoding METAR data can be a time-consuming task; however, METARs are used for numerous applications in the NWS, Federal Aviation Administration (FAA), and some private forecasting companies. Although METARs are encoded, transmitted, and decoded by computers today, the ability to decode a METAR still is a good skill to learn, especially to analyze international data.

The following METAR is decoded section-by-section to provide an example of the decoding process. For more information about METARs, consult the NWS web page at http://weather.noaa.gov.

KIAD 081351Z 18012KT 3SM +SN OVC021 02/M04 A2994 RMK A02 SNB23 SLP998 P0045 T00181036=

KIAD
The METAR begins with the *four-letter ID* for the station. Common prefixes for the IDs include the following:

- C   Canada
- K   Continental United States
- M   Mexico, Central America, and portions of the Caribbean (Bahamas, Cuba, Haiti, Dominican Republic, Jamaica)
- P   Alaska, Hawaii, Guam, and other Pacific Islands
- T   Other Caribbean islands, including Puerto Rico and the Virgin Islands

In this case, **K** means the contiguous United States and **IAD** refers to Dulles International Airport in Washington, DC.

081351Z
The *date and time of the observation* follow the station ID. The first two numbers are the date of the month, the next four numbers are the time, and **Z** represents Zulu time, or UTC. In this case, it is the 8th day of the month (e.g., March 8th) at 1351 UTC.

18012KT
The *wind direction* (in degrees) *and wind speed* (in knots) precede the letters **KT** (knots). As in all meteorological applications, the wind direction is the direction from which the wind is blowing. For example, a 180° wind is from the south and a 325° wind is from the northwest. For this METAR report, the wind is from the south (180°) at 12 knots.

Figure 11 –Orientation of Wind in Degrees

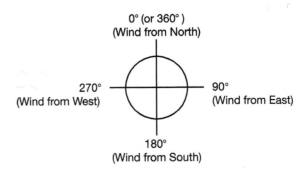

3SM
The next section describes the *visibility* in statute miles (**SM**) and fractions of a statute mile. Values less than 1/4 mile are indicated as M1/4. In other countries, visibility is given in meters. In this METAR report, the visibility is 3 statute miles.

+SN
The *current weather* at the time of the measurement follows the visibility. Table 1 lists abbreviations and acronyms that are used for current weather conditions. In this example, the + indicates heavy intensity and the **SN** stands for snow. Thus, heavy snow was observed at 1351 UTC (the time of the observation). It is possible to have multiple weather types in the same observation (e.g., –TSRA BR means there is a light thunderstorm with rain and mist).

OVC021
This next term is the *cloud cover* and *height of the cloud base* (in hundreds of feet). The cloud cover abbreviations also appear in Table 1. Multiple entries may occur for this term because it is possible to have more than one cloud deck (e.g., **SCT060 OVC100** translates to scattered clouds at 6000 feet and overcast skies at 10,000 feet). For the example METAR at KIAD, the skies are overcast at 2100 feet.

02/M04
The *temperature* and *dew point* in degrees Celsius appear next. If a temperature is negative, an **M** is inserted before the value. In this example, the temperature is 2°C and the dew point is –4°C.

A2994

The **altimeter setting** follows the temperature and dew point. The altimeter setting is reported in either inches of mercury (**A**) or millibars (**Q**). In this case, the altimeter setting is 29.94 inches Hg (i.e., inches of mercury).

RMK

The next section contains *remarks* – details that may be of particular importance. The remarks may list the timing and amount of precipitation, the sea-level pressure, a more precise reading of temperature and dew point, and comments by NWS personnel. The remarks section is not required for a METAR to be complete. The following remarks are common to METARs.

A02

This remark indicates that the station is fully automated with a sensor that can discriminate between different types of precipitation.

SNB23

This remark means that snow began at 23 minutes past the hour — that is, at 1323 UTC.

SLP998

A remark starting with **SLP** indicates the *sea-level pressure* in tenths of a millibar. The first number (either a 9 or a 10) has been dropped. Either the computer decoder or the human interpreter decides which number is appropriate to add. Because sea-level pressures range from 970 mb to 1030 mb, the missing 9 or 10 can be determined unambiguously. In this case, the sea-level pressure is 999.8 mb.

P0045

This remark represents the *precipitation* in hundredths of an inch that fell during the last hour. P0000 stands for a trace of precipitation. At this station, 0.45 inches of precipitation was measured during the last hour. For snow, the precipitation amount listed the liquid water equivalent (i.e., the amount that would be measured if the snow were melted). According to this report, snow had been falling and the liquid water equivalent would measure 0.45 inches.

T00181036

The final remark for this example provides a more precise version of the *air temperature* and *dew point* (degrees Celsius and tenths of a degree) than provided earlier. The first four numbers represent the temperature and the second four numbers represent the dew point. In each set of four numbers, the first number indicates whether the following temperature is above (0) or below (1) zero degrees Celsius. In this example, the temperature is 1.8°C and the dew point is –3.6°C.

=

The equal sign denotes the *end of the METAR report*. It allows observers to recognize whether all of the information was transmitted or if some information was lost.

Table 1 – METAR Abbreviations

Code	Meaning	Code	Meaning
+	heavy intensity	OHD	overhead
(no symbol)	moderate intensity	OVC	overcast (8/8 coverage)
-	light intensity	OVR	over
ACC	altocumulus castellans	PCPN	precipitation
A02	automated w/ precipitation discriminator	PE/PL	ice pellets
AO1	automated w/o precipitation discriminator	PK WND	peak wind
AUTO	fully automated report	PNO	precipitation amount N/A
B	began	PO	dust/sand whirls
BKN	broken clouds (5/8-7/8 coverage)	PRES	pressure
BL	blowing	PRESFR	pressure falling rapidly
BR	mist	PRESRR	pressure rising rapidly
CA	cloud-air lightning	PY	spray
CB	cumulonimbus cloud	RA	rain
CBMAM	cumulonimbus mammatus	RVR	runway visual range
CC	cloud-cloud lightning	S	south
CG	cloud-ground lightning	SA	sand
CIG	ceiling	SCT	scattered clouds (3/8-4/8 coverage)
CLR	clear (no clouds)	SE	southeast
CONS	continuous	SFC	surface
DR	drifting	SG	snow grains
DS	dust storm	SH	shower
DSIPTG	dissipating	SKC	sky clear
DSNT	distant	SLP	sea level pressure
DU	widespread dust	SLPNO	sea level pressure N/A
DZ	drizzle	SM	statute miles
E	east or ended	SN	snow
FC	funnel cloud	SNINCR	snow increasing rapidly
FEW	few clouds (0 - 2/8 coverage)	SP	snow pellets
FG	fog	SQ	squall
FRQ	frequent	SS	sandstorm
FROPA	frontal passage	SW	snow shower or southwest
FT	feet	TCU	towering cumulus
FU	smoke	TS	thunderstorm
FZ	freezing	TSNO	thunderstorm info N/A
G	gust	TWR	tower
GR	hail	UNKN	unknown
GS	small hail or snow pellets	UP	unknown precipitation
HZ	haze	UTC	Universal Time Code
IC	ice crystals or in-cloud lightning	V	variable
INCRG	increasing	VC	in the vicinity
INTMT	intermittent	VIS	visibility
KT	knots	VR	visual range
LTG	lightning	VRB	variable
M	minus, less than	VV	vertical visibility
MOV	moving	W	west
N	north	WND	wind
NE	northeast	WSHFT	wind shift
NW	northwest	Z	Zulu Time
OCNL	occasional		

12. **(Advanced Students/Meteorology Majors)** Using Table 2 and the METAR decoding guide, decode the METAR observation at 0019 UTC from May. Note: KTIK is in the Central Time zone and observes Daylight Savings Time, when appropriate.

Station _____   Temperature (highest precision) _____ °C

Date _____   Dew Point (highest precision) _____ °C

Time (local time) _____   Visibility _____ statute miles

Altimeter _____ in Hg   Precipitation _____ inches during last hour

Wind Direction _____ degrees   Cardinal Wind Direction _____

Wind Speed _____ knots

Current Weather Conditions: _____

_____

Cloud Coverage and Heights: _____

_____

Remarks Section: _____

_____

_____

Table 2 – METAR Data from May 1999

KTIK 032255Z 16013G20KT 7SM FEW025 SCT040CB OVC250 22/20 A2957 RMK SLP005 CB 30SW MOV NE CBMAM OHD 8/303 9/305=
KTIK 032332Z 16011G20KT 6SM -TSRA FEW023 SCT040CB OVC250 22/19 A2959 RMK TS 15SW MOV NE=
KTIK 032355Z 18012KT 6SM -TSRA FEW023 SCT040CB OVC250 21/20 A2962 RMK OCNL LTGICCCCG TS 3SW MOV NE PRESRR SLP023 60001 8/903 9/404 55000=
KTIK 040001Z 17007KT 2SM -TSRA BR FEW025 BKN040CB OVC250 21/20 A2962 RMK OCNL LTGICCCCG TS 3SW MOV NE=
KTIK 040003Z 17006KT 2SM -TSRA BR BKN020CB BKN080 OVC250 21/20 A2961 RMK OCNL LTGICCCCG TS OHD MOV NE=
KTIK 040010Z RMK TORNADO 13SW MOV NE. LARGE TORNADO ON THE GROUND TAKE COVER IMMEDIATELY.=
KTIK 040019Z 08008KT 1SM +FC TSRAGR OVC020CB 21/20 A2955 RMK TORNADO 9SW MOV NE FRQ LTGICCCCG TS OHD MOV NE GR 1 1/4 PRESFR=
KTIK 040031Z 08006KT 6SM +FC OVC020CB 21/20 A2956 RMK TORNADO 5SW MOV NE FRQ LTGICCCCG TS OHD MOV NE PRESRR=
KTIK 040036Z 08006KT 6SM +FC OVC020CB 21/20 A2954 RMK TORNADO 3SW MOV NE FRQ LTGICCCCG TS OHD MOV NE PRES UNSTDY=
KTIK 040043Z COR RMK TORNADO 1W MOV NE. EVACUATING STATION=
KTIK 040043Z RMK TORNADO OHD. EVACUATING STATION=

13. **(Advanced Students/Meteorology Majors)** Using the METAR decoding guide, decode the following METAR observation from September. Note: KNEW is in the Central Time zone and observes Daylight Savings Time, when appropriate.

METAR KNEW 251353Z 06025KT 1 1/4SM +RA BR FEW001 BKN011 OVC023 24/23 A2973 RMK AO2 PK WND 06032/1338 SLP065 P0004 T02440228

Station _____   Temperature (highest precision) _____ °C
Date _____   Dew Point (highest precision) _____ °C
Time (local time) _____   Visibility _____ statute miles
Altimeter _____ in Hg   Precipitation _____ inches during last hour
Wind Direction _____ degrees   Cardinal Wind Direction _____
Wind Speed _____ knots

Current Weather Conditions: _____

_____

Cloud Coverage and Heights: _____

_____

What does "**PK WND 06032/1338**" mean? _____

_____

Remaining Remarks: _____

_____

_____

14. **(Advanced Students/Meteorology Majors)** Completely encode the following report *in the correct order*.

ASOS – augmented observation from Wilmington, NC (ILM) at 3:26 PM EDT on the 24th:
Temperature 23°C; dew point 23°C; altimeter setting 29.87" of Hg; sky cover is 3/8 at 800 feet, 5/8 at 2300 feet, and 7/8 at 3000 feet, visibility 3 miles with a thunderstorm, rain, and mist. Winds from the south at 13 knots, with gusts to 22 knots.

_____

_____

15. **(Advanced Students/Meteorology Majors)** Plot the METAR reports from Table 3 on their corresponding circle on Figure 12. Use the station model format as shown in Figures 1–5. The surface station locations are provided in Figure 6. When indicating sky cover on the station model, use the greatest amount of cloud cover. For example, if a METAR statement displays the cloud cover as **FEW021 BKN 050**, the amount of sky covered should be displayed as broken on the station model. You also may check to see if your plots are consistent with the station model plots shown in Figure 7.

Table 3 – METAR Data

KAIG 170315Z AUTO 17004KT 3SM +RA FG SCT010 BKN018 OVC030 14/14 A2980 RMK AO2=
KABR 170256Z 04010KT 10SM -RA BKN050 OVC060 06/04 A2978 RMK AO2 SLP089 P0000 60004 T00610044 58021=
KCPR 170255Z 32010KT 1 1/4SM -SN BR SCT008 OVC027 00/00 A2992 RMK AO2 SLP145 P0001 60010 T00000000 51023 $=
KLIT 170250Z 18007KT 5SM HZ SKC 19/15 A2991 RMK SLP129 55000=
KAEL 170312Z AUTO 14008KT 5SM FG SCT016 BKN037 OVC044 17/16 A2962 RMK AO2=
KAFN 170252Z AUTO 18005KT 3SM -RA BR FEW001 BKN055 OVC090 13/12 A2991 RMK AO2 SLP135 P0005 60006 T01330122 53013 TSNO=
KOKC 170256Z 17015KT 10SM CLR 22/17 A2973 RMK AO2 SLP056 T02170172 53005=
KLBF 170256Z 34008KT 8SM -TSRA OVC060 06/03 A2975 RMK AO2 PK WND 33030/0228 TSE14B28RAE09B32 PRESRR SLP071 P0002 60007T00610028 51056=

Table 4 – Station ID, Station Name, State, Latitude, and Longitude for METAR Observations Listed in Table 3

Station ID	Name	State	Latitude	Longitude
KAIG	Antigo Automatic Weather Observing System	WI	45.03° N	89.14° W
KABR	Aberdeen Regional Airport	SD	45.45° N	98.43° W
KAEL	Albert Lea Automatic Weather Observing System	MN	43.68° N	93.37° W
KAFN	Jaffrey Municipal Airport	NH	42.80° N	72.00° W
KCPR	Casper (Natrona County International Airport)	WY	42.92° N	106.47° W
KLBF	North Platte Regional Airport	NE	41.13° N	100.68° W
KLIT	Little Rock (Adams Field)	AR	34.73° N	92.23° W
KOKC	Oklahoma City (Will Rogers World Airport)	OK	35.40° N	97.60° W

Figure 12 – Station Models for METAR Decoding Exercise

Radiosondes and Soundings

LAB ACTIVITY OBJECTIVES:

- Given a sounding plotted on a thermodynamic diagram, you will read data values from the diagram, calculate relative humidity at different levels, and locate the tropopause, lowest freezing level, and any inversion layers.
- Given soundings from the same station on two consecutive days, you will describe changes in the atmosphere from one day to the next.
- Given data from an atmospheric sounding, you will plot upper-air observations on a blank thermodynamic diagram.
- The advanced student will describe how height data are calculated from the data measured by a radiosonde.

MATERIALS NEEDED:

- Laboratory manual
- Pencil or pen
- Colored pencils
- Ruler

GLOSSARY:

Inversion	Saturation Mixing Ratio
Isobar	Sounding
Isotherm	Thermodynamic Diagram
Mixing Ratio	Thickness
Radiosonde	Tropopause
Relative Humidity	Vapor Pressure

BACKGROUND:

The large high- and low-pressure systems that most influence our daily weather extend well above the earth's surface. Weather offices worldwide use **radiosondes** to take measurements of the atmosphere above the surface and use these data for forecasts. The radiosonde is an instrument platform with meteorological sensors and a radio transmitter that is carried aloft by a balloon. By international agreement, operational and routine ascents occur from select observation stations worldwide at just prior to 0000 UTC and 1200 UTC. The radiosonde instruments measure air temperature, humidity, and pressure as the balloon carries them to altitudes of approximately 30 km (~10 mb). The transmitter transmits the data to a ground-based receiver. This receiver tracks the path of the balloon, allowing for the calculation of wind direction and wind speed. Historically, the set of observations measured by a single radiosonde is called a **sounding**.

The main component of the radiosonde is a sturdy, lightweight, white, cardboard instrument package that is approximately the size of a box of breakfast cereal. The radiosonde package includes a thermistor, a hygristor, an aneroid barometer, and a transmitter. The temperature sensor is a resistance thermistor that measures the change in electrical resistance as the air temperature changes. The hygristor is a humidity sensor consisting of a glass slide covered with a moisture sensitive film of lithium chloride (LiCl). The electrical resistance of lithium chloride changes as the atmospheric humidity changes. Pressure is measured using an aneroid barometer. As the radiosonde ascends, the volume of the aneroid canister expands in response to a reduction in the atmospheric pressure aloft. The radio transmitter operates on a modulated frequency of 1680 MHz.

A balloon carries the radiosonde aloft. The balloon is inflated initially with hydrogen to an approximate diameter of 2 meters (6 feet). As the balloon ascends, it expands in size to a diameter between 8 and 11 meters before it bursts at about 30 km. An attached parachute returns the package safely to the ground. Those radiosondes that are found can be returned to the National Weather Service and refurbished for subsequent flights, saving a considerable amount over the cost of a new radiosonde.

During its ascent, the radiosonde transmits temperature and **relative humidity** data at 10 mb pressure increments. The altitudes of these levels are calculated using an equation that relates the vertical **thickness** of a layer to the mean layer temperature, the humidity of the layer, and the air pressure at the top and bottom of the layer (i.e., the hypsometric equation).

After the radiosonde has concluded its flight, the data are encoded and transmitted to the National Centers for Environmental Prediction (NCEP). The data are processed for analysis on upper air charts and for use in numerical weather prediction models.

The temperature, humidity, and wind data are encoded for *significant* and *mandatory* levels only. By international convention, the *mandatory levels* are the surface, 1000 mb, 925 mb, 850 mb, 700 mb, 500 mb, 400 mb, 300 mb, 250 mb, 200 mb, 150 mb, 100 mb, 70 mb, 50 mb, and 10 mb. *Significant pressure levels* are those altitudes where the sensors detect a significant change in the vertical profile of temperature or humidity.

THE SKEW–T LOG–P DIAGRAM

To better understand the vertical structure of the atmosphere, meteorologists most commonly plot soundings using a type of diagram called a Skew–T Log–P (or Skew–T for short). The Skew–T is a **thermodynamic diagram** because it relates thermodynamic variables like pressure, temperature, and moisture to one another. Meteorologists use the Skew–T diagram to examine graphically how air parcels interact with their environment. For example, the diagram can be used to determine if air parcels at a given altitude will rise or sink if slightly disturbed.

One especially useful characteristic of the Skew–T is that area on the diagram is proportional to energy. Hence, the amount of energy that a thunderstorm could tap can be calculated and visualized easily using a Skew–T.

Pressure (Isobars) on a Skew–T

At first glance, Skew-T diagrams appear complicated, but they become easy to use with a little practice. To make the diagram as useful as possible, the design of the diagram reflects many realistic characteristics of the atmosphere. For example, because pressure decreases logarithmically with height, the vertical coordinate of the Skew–T is pressure plotted on a logarithmic scale (see Figure 1). Therefore, 1000 mb is at the bottom of diagram (near or below the surface), and 100 mb is at the top of the diagram. Table 1 displays how some of these pressure levels (**isobars**) relate to altitude. The exact relationship between pressure and altitude varies daily, so the values in Table 1 are approximate.

Temperature (Isotherms) on a Skew–T

The horizontal coordinate on the Skew-T is temperature (Figure 2). The temperature values increase towards the bottom and right of the diagram. The temperature grid lines (in red) are displayed in degrees Celsius. On Figure 2, the shaded area represents temperatures above freezing (i.e., T > 0°C). Because the **isotherms** are skewed to the right with lower pressure, the diagram is called the "Skew–T Log–P."

Mixing Ratio on a Skew–T

Another set of lines on the Skew–T diagram represent the moisture content of the air. The purple, dashed lines on Figure 3 depict the **mixing ratio**. Mixing ratio (w) is defined as the mass of water vapor (in grams) divided by the mass of *dry air* (in kilograms). Like **vapor pressure**, the mixing ratio is a function of dew point and is a measure of the actual amount of water vapor in the air. Also, like saturation vapor pressure, the **saturation mixing ratio** (w_s) is a function of temperature and represents the maximum amount of water vapor that can be present at a given temperature. On Figure 3, the mixing ratio lines slant towards the upper right of the diagram and are labeled near the top of the line.

Figure 1 – Pressure Lines (Isobars) on a Skew-T Diagram

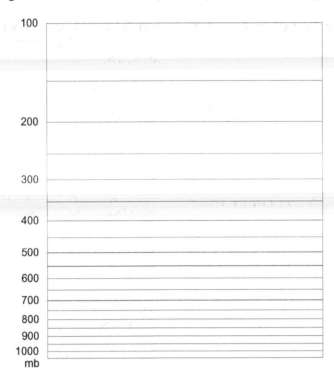

Table 1 – Approximate Relationship Between Pressure Levels and Altitude

Pressure Level	Approximate Height
1000 mb	Surface
850 mb	5,000 ft
500 mb	18,000 ft
300 mb	30,000 ft

Figure 2 – Skew-T Diagram with Pressure (Isobars) and Temperature (Isotherms) Lines

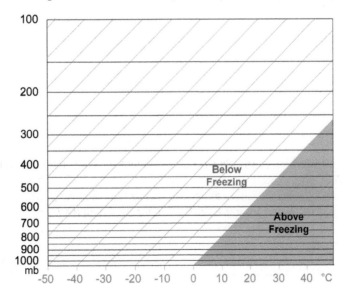

Figure 3 – Basic Skew-T Diagram with Isobars, Isotherms, and Mixing Ratio Lines

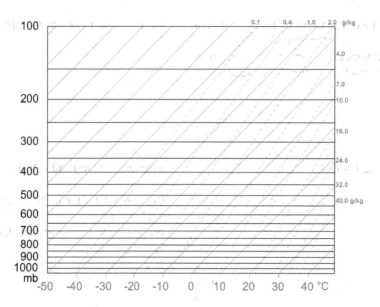

Figure 4 – Skew-T Diagram of the Sounding Observations Listed in Table 2

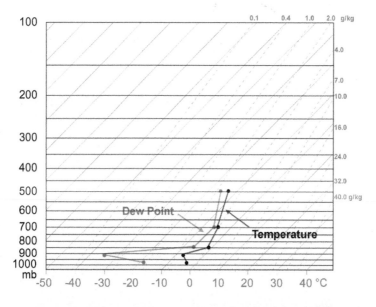

Table 2 – Selected Observations from a Greensboro, NC Sounding from 1200 UTC on 4 December 2002

GSO: Greensboro, NC		1200 UTC on 4 Dec 2002	
Pressure (in mb)	Height above sea level (in m)	Temperature (in °C)	Dew Point (in °C)
999	270	−2.9	−18.9
925	877	−6.9	−34.9
850	1545	−0.5	−4.2
700	3094	−5.3	−5.6
500	5700	−13.7	−15.1

16. **(Advanced Students/Meteorology Majors)** Both sets of sounding data in this lab contain height data along with the pressure, temperature, moisture, and wind information. The radiosonde instrument package, however, does not measure height. Describe how the height data are obtained.

height is obtained by the Dropp in pre

17. **(Advanced Students/Meteorology Majors)** Atmospheric thickness is defined as the difference in height between two pressure levels. Using Tables 2 and 3, calculate the 1000–500 mb thickness values for both soundings. How do the thickness values differ between 16 April 2002 and 27 January 2000? Using concepts discussed in lecture or your textbook, what is the physical reason as to why the two thickness values differ?

1000–500 mb thickness for ABR on 16 April 2002 at 1200 UTC _____ meters

1000–500 mb thickness for FWD on 27 January 2000 at 1200 UTC _____ meters

18. **(Advanced Students/Meteorology Majors)** What is the physical reason for the difference in tropopause heights between the two soundings?

Soundings and Stability

LAB ACTIVITY OBJECTIVES:

- You will examine and discuss the relationship between dry adiabats and moist adiabats on a sounding.
- Given a plotted sounding, you will estimate graphically the lifting condensation level, convective condensation level, and equilibrium level.
- Using a plotted sounding and associated tabular data, you will calculate four different stability indices and use their values to discuss the likelihood of thunderstorms and severe thunderstorms for both warm and cold seasons.
- The advanced student will discuss pros and cons of using a specific stability index in a given situation.
- The advanced student will discuss how the timing of operational radiosonde launches influences the calculation of stability indices.

MATERIALS NEEDED:

- Laboratory manual
- Pencil or pen
- Colored pencils

GLOSSARY:

Adiabat	K-Index	Stability Indices
Adiabatic	Lapse Rate	Stable
Air Parcel	Lifted Index	SWEAT Index
Buoyant	Lifting Condensation Level (LCL)	Total-Totals Index
Convective Condensation Level (CCL)	Mixing Ratio	Unstable
Dry Adiabatic Lapse Rate	Moist Adiabatic Lapse Rate	Vertical Wind Profile
Environmental Lapse Rate	Radiosonde	
Equilibrium Level (EL)	Saturation Mixing Ratio	
Ideal Gas Law	Sounding	
Instability	Stability	

BACKGROUND:

The **stability**, or **instability**, of the atmosphere has major implications for the amount of vertical motion that may occur in the atmosphere. Vertical motion is a key ingredient in most atmospheric processes, from the development of clouds and precipitation to the transport and dispersion of chemical constituents.

At a given time and place, the stability of the atmosphere depends on the vertical profiles of both temperature and dew point. These profiles, typically measured using **radiosondes**, depict how the temperature and moisture in the atmosphere change with altitude over a given location.

One method to determine the atmospheric stability is the "parcel method." The parcel method assumes that **air parcels** follow theoretical principles of thermodynamics. That is, as the pressure of the air parcel changes, the parcel responds according to the **ideal gas law** and other physical laws. The parcel method compares the behavior of a rising parcel with the temperature and moisture conditions of its environment, as measured by the radiosonde.

The rate at which temperature changes with height in the atmosphere is called the **lapse rate**. The parcel method requires three lapse rates: (1) the **dry adiabatic lapse rate** for an unsaturated parcel, (2) the **moist adiabatic lapse** rate for a saturated parcel, and (3) the **environmental lapse rate** (i.e., the temperature profile as *measured* by the radiosonde). The parcel method assumes that the air parcel remains distinct from and does not mix with its environment. Thermodynamically stated, the air parcel does not transfer heat between itself and its environment; that is, changes to the state (e.g., temperature, pressure, volume, and moisture content) of the parcel are **adiabatic**. If adiabatic changes to the parcel result in it becoming more **buoyant** than its surrounding environment, the parcel is **unstable** and will rise. Conversely, if small vertical displacements to the parcel result in the parcel becoming less buoyant, then it is stable and returns to its initial position. To apply the parcel method, the Skew–T Log-P (hereafter called Skew–T for short) diagram will be used.

DRY–ADIABATIC LAPSE RATE

An unsaturated air parcel (i.e., containing no liquid water) cools at a rate of 10°C for every 1000 meters it rises in the atmosphere. Similarly, the parcel warms 10°C per 1000 m as it sinks. This rate of temperature change per kilometer is called the **dry adiabatic lapse rate**. It is the rate that the temperature of an unsaturated air parcel changes solely as a result of the expansion or compression associated with the parcel's ascent or descent. The expansion and compression of air parcels result from the change of pressure with height (Remember: pressure decreases with height), as air obeys the ideal gas law.

Skew-T diagrams include lines that visually depict the dry adiabatic lapse rate. These curved lines, called dry **adiabats**, slant upward and to the left on Skew-T diagrams (solid, blue lines on Figure 1). Dry adiabats are used to examine the changes in temperature of an unsaturated air parcel as it undergoes small vertical displacements. For example, assume that an unsaturated air parcel at 1000 mb initially has a temperature of 18°C. If it were lifted to 800 mb and remained unsaturated along its rise, it would have a temperature of approximately 0°C when it stopped at 800 mb. To visualize this process using the Skew–T diagram, first locate this parcel at intersection of 1000 mb and 18°C. Represent the ascent of the parcel by progressing upward and to the left, parallel to the dry adiabat (solid, blue line), from 1000 mb to 800 mb. To find the final temperature, read the value of the isotherm (solid, red line) at 800 mb.

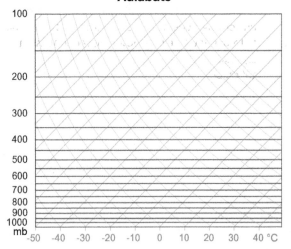

Figure 1 – Skew–T Diagram with Dry Adiabats

MOIST–ADIABATIC LAPSE RATE

A saturated air parcel (e.g., parcel in a cloud) cools (or warms) at an average rate of ~6°C for every 1000 meters it rises (or sinks) in the atmosphere. This moist adiabatic lapse rate is less than its dry adiabatic counterpart because it includes the effect of condensation and evaporation of water vapor. The moist adiabatic lapse rate starts with the dry adiabatic lapse rate (i.e., the temperature change resulting solely from the expansion or compression of the parcel) and adds the release (or absorption) of latent heat resulting from the condensation (or evaporation) of water vapor. Hence, a rising saturated parcel cools from expansion (at 10°C per km) and warms from condensation (at ~4°C per km) for a total cooling rate of ~6°C per km. Similarly, a sinking saturated parcel warms from compression (at 10°C per km) and cools from evaporation (at ~4°C per km).

On the Skew-T diagram, moist adiabats (also called saturation adiabats) are used to examine the changes in temperature of a saturated air parcel as it undergoes small vertical displacements. These curved lines appear as dashed, green lines on Figure 2. For example, assume that a saturated air parcel at 900 mb has an initial temperature of 20°C. Following the moist adiabat (dashed green line) to 600 mb reveals that the air parcel would have a temperature of –5°C at 600 mb.

Most Skew-T diagrams (e.g., Figure 3) overlay isobars, isotherms, lines of constant **saturation mixing ratio**, dry adiabats, and moist adiabats. Additionally, wind barbs usually are plotted on the right side of the diagram as a visual representation of the **vertical wind profile** (Figure 3). Unfortunately, no standard for colors or line types has been developed and universally accepted. Hence, it may be difficult at first to use the Skew-T. However, recognizing and visually filtering the lines based upon their shape and orientation becomes easier with practice.

Figure 2 – Skew–T Diagram with Moist Adiabats

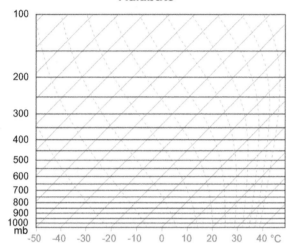

LIFTING CONDENSATION LEVEL

The parcel method normally begins with an unsaturated parcel that becomes saturated as its temperature cools to its dew point. Hence, the parcel cools at the dry adiabatic lapse rate until it becomes saturated. At that point, the parcel cools at the moist adiabatic lapse rate if it remains unstable and continues to rise. The level where an unsaturated parcel must be lifted to become saturated is called the **lifting condensation level**, or **LCL**. Below the LCL, the parcel cools (or warms) dry adiabatically; above the LCL, the parcel cools (or warms) moist adiabatically.

ATMOSPHERIC INSTABILITY

The atmosphere is unstable when an air parcel is more buoyant than its environment. As a result, the air parcel overcomes the force of gravity and rises in the atmosphere. To envision this instability, imagine that the air parcel is a beach ball that is inflated with air, and imagine that the atmosphere is an ocean. If you were to push the beach ball downward into the ocean and then let go of the ball, it would move upward quickly because it was less dense (more buoyant) than the water.

In an unstable atmosphere, a parcel of rising air is warmer and, thus, less dense than the environmental air (e.g., temperature on the **sounding**) at the same pressure level. For an unsaturated parcel to be unstable, the environmental lapse rate (as measured by the radiosonde) must exceed the dry adiabatic lapse rate. For a saturated parcel to be unstable, the environmental lapse rate must exceed the moist adiabatic lapse rate. An unstable atmosphere supports the formation of clouds and, in some cases, severe thunderstorms. Cumulus clouds are signs of atmospheric instability.

Conversely, the atmosphere is **stable** when the air parcel is less buoyant than its environment. As a result, the parcel will descend in the atmosphere. Although clouds can form in a stable atmosphere, they do not result from buoyant parcels of air. Instead, air parcels must be forced upward by some other mechanism (e.g., forced over terrain, forced over a vertically sloped frontal boundary) than buoyancy. Stratus clouds typically are signs of atmospheric stability.

Only rarely does the environmental lapse rate exceed the dry adiabatic lapse rate. When it occurs, it is measured during the daytime near the surface, where solar heating quickly warms the ground. It is relatively common, however, to measure an environmental lapse rate greater than the moist adiabatic lapse rate. Hence, for an air parcel to become more buoyant than the environment, it typically must become saturated first. Meteorologists examine soundings to determine if there is an opportunity for a parcel to attain saturation and thereafter become unstable.

Figure 3 – Skew–T Diagram with Dry and Moist Adiabats

Date & Time: _____

Station: _____

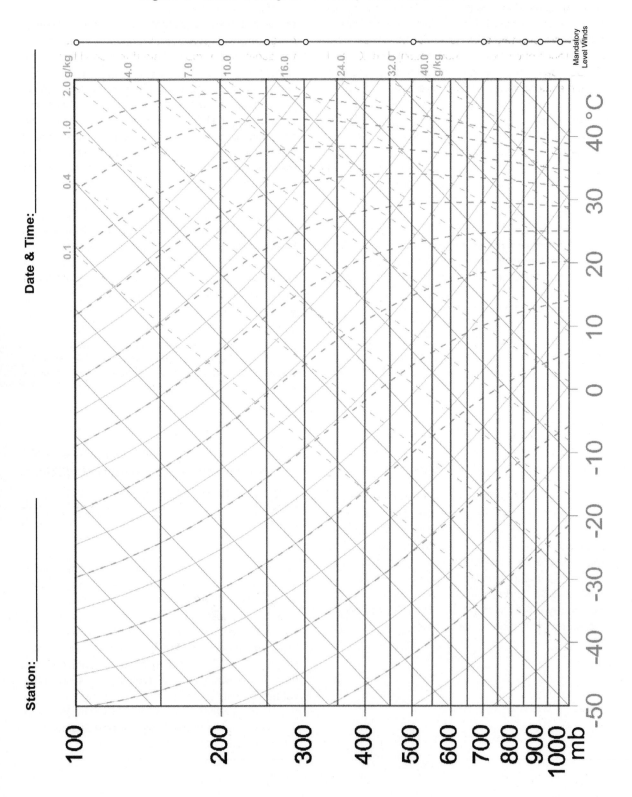

The level where the parcel first becomes saturated is either the lifting condensation level (LCL), if the parcel has been mechanically lifted (e.g., by a front), or the **convective condensation level (CCL)**, if the parcel rises through buoyancy (i.e., convection). The LCL and the CCL define the altitude of cloud bases that result from the rising motion. Both levels are calculated from soundings using the process demonstrated in Figure 4. Upon reaching the LCL or CCL, a parcel continues to rise if the environmental lapse rate exceeds the moist adiabatic lapse rate. [In other words, the temperature of the parcel (as indicated by its moist adiabat) is warmer than the actual temperature measured by the radiosonde so that the moist adiabat of the unstable parcel is located to the right of the observed temperature curve on the Skew-T diagram.] The parcel may rise throughout most of the depth of the troposphere if this condition is met throughout that depth. At some level, however, the environment will become warmer than the parcel and the parcel will stop ascending. This level is called the **equilibrium level (EL)**. The EL also can be calculated from soundings, using the process demonstrated in Figure 5. At the EL, the parcel's moist adiabat crosses the observed temperature trace. The steps used to calculate the LCL, CCL, and EL are outlined below.

Steps used to calculate the LCL (Figure 4):

1. From the dew point curve (T_D) at the given pressure, follow the **mixing ratio** line upwards. [This process physically represents lifting the parcel while maintaining the same number of water vapor molecules (i.e., constant mixing ratio) — no condensation.]
2. From the temperature curve (T) at the given pressure, follow the dry adiabat upwards. [This process physically represents the cooling of the unsaturated parcel as it expands (i.e., moves into lower pressure).]
3. The intersection of these lines is the lifting condensation level (LCL).

Step used to calculate the CCL (Figure 4):

1. From the dew point curve (T_D) at the given pressure, follow the mixing ratio line upwards until it intersects the environmental temperature curve. [This process determines the temperature for saturation to occur for the given moisture content of an air parcel.]

Step used to calculate the EL (Figure 5):

1. Find the CCL.
2. Follow the moist adiabat upwards from the CCL until it intersects the environmental temperature curve. [This process represents the rise of an unstable air parcel until it becomes cooler than its environment.]

Figure 4 – Schematic Diagram Demonstrating How to Compute the Lifting Condensation Level (LCL) and the Convective Condensation Level (CCL)

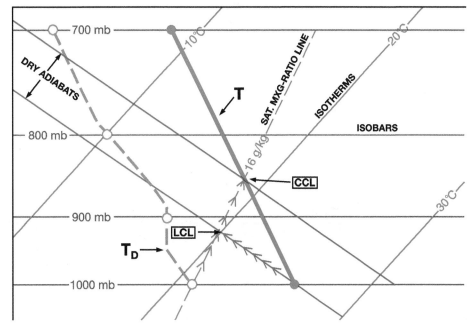

Figure 5 – Schematic Diagram Demonstrating How to Compute the Equilibrium Level (EL)

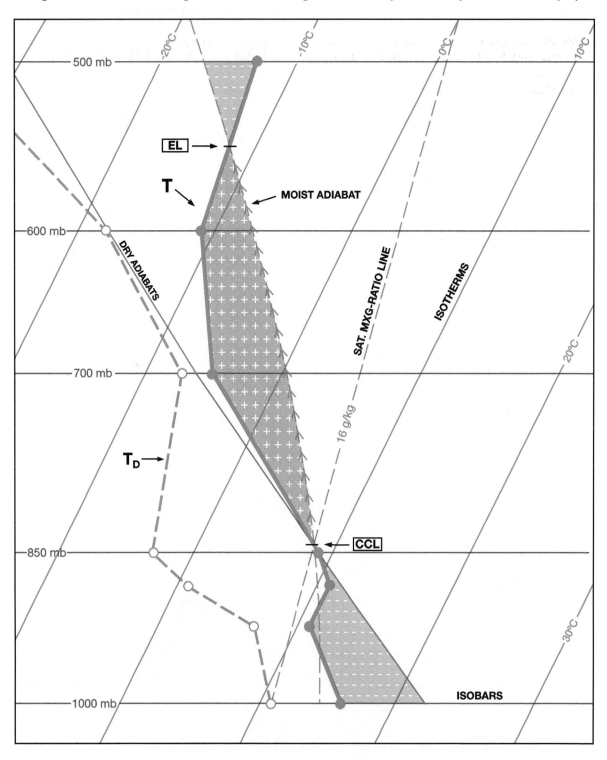

STABILITY INDICES

Because of the detrimental effects of thunderstorms on life and property, meteorologists have developed numerous **stability indices** to assess instability and thunderstorm potential. These indices are empirical formulas; hence, they are not mathematical expressions of any laws of physics. Instead, they have resulted from observations and experimentation, as forecasters noted relationships between sounding data and thunderstorm activity. Meteorologists use stability indices to provide a *quick* estimate of the potential for thunderstorm activity. Four of these indices – the **lifted index**, **K-Index**, **Total-Totals Index**, and **SWEAT Index** – are the most common in thunderstorm forecasting.

Lifted Index (LI)

The lifted index (LI) provides an estimate of how unstable an air parcel would be if the parcel were lifted from the surface to 500 mb. The lifted index is defined as the difference of the air parcel's temperature at 500 mb and the environmental temperature at 500 mb. To calculate the lifted index, complete the following steps:

1. Determine the lifting condensation level (LCL) using the procedure described above (e.g., Figure 4).
2. From the LCL, follow the moist adiabat upward to the 500-mb level.
3. Determine the temperature (in °C) of the lifted parcel at 500 mb ($T_{P\,500}$). To do this, simply read the temperature of the moist adiabat where it crosses the 500-mb isobar.
4. Subtract the lifted parcel temperature (in °C) at 500 mb ($T_{P\,500}$) from the environmental temperature (T_{500}) at 500 mb, as indicated in the equation below.

$$LI = T_{500} - T_{P\,500}$$

Physically, the LI indicates that the amount of instability in the middle of the troposphere is proportional to the temperature difference between the lifted parcel and its environment at 500 mb. If the lifted parcel is warmer than its environment at 500 mb (i.e., unstable), then the lifted index is negative. If the lifted parcel is cooler than its environment at 500 mb (i.e., stable), then the lifted index is positive. Table 1 provides an interpretation of the values for the lifted index.

Table 1 – Interpretation of Values for the Lifted Index

Lifted Index	Interpretation
0 or greater	Stable conditions
–1 to –4	Marginal instability
–5 to –7	Large instability
–8 or less	Extreme instability

K-Index (K)

The K–Index was designed to be an indicator of non-severe thunderstorms. Calculation of the K-Index (K) is determined using the equation below and values observed directly by the radiosonde.

$$K = (T_{850} - T_{500}) + T_{D\,850} - (T_{700} - T_{D\,700}) \,,$$

where T_{850} = environmental temperature at 850 mb (in °C),

T_{500} = environmental temperature at 500 mb (in °C),

$T_{D\,850}$ = environmental dew point at 850 mb (in °C),

T_{700} = environmental temperature at 700 mb (in °C), and

$T_{D\,700}$ = environmental dew point at 700 mb (in °C).

Physically, the first term in the K–Index is a measure of the environmental lapse rate between 850 mb and 500 mb. The second and third terms are measures of the moisture available for convection. The K–Index is a poor indicator of severe thunderstorms because dry air at 700 mb enhances convective instability (an important ingredient for severe thunderstorm development), yet the resulting K–Index value is low. Table 2 provides an interpretation of the values for the K–Index.

Table 2 – Interpretation of Values for the K–Index

K-Index	Interpretation
less than 15	Convection not likely
15 to 25	Small potential for convection
26 to 39	Moderate potential for convection
40 or greater	High potential for convection

Total Totals Index (TT)

The Total Totals Index, like the K-Index, is a measure of the environmental lapse rate and the moisture available for convection. The Total Totals Index is a combination of the Vertical Totals ($VT = T_{850} - T_{500}$) and the Cross Totals ($CT = T_{D\,850} - T_{500}$). The sum of these two values is the Total Totals (TT). This index has proven to be useful in diagnosing regions of severe thunderstorms.

$$TT = T_{850} + T_{D\,850} - 2 \times T_{500}\,,$$

where T_{850} = environmental temperature at 850 mb (in °C), $T_{D\,850}$ = environmental dew point at 850 mb (in °C), and T_{500} = environmental temperature at 500 mb (in °C).

Table 3 provides an interpretation of the values for the Total Totals Index.

Table 3 – Interpretation of Values for the Total Totals Index

Total Totals Index	Interpretation
less than 44	Convection not likely
44 to 50	Convection likely
51 to 52	Isolated severe storms
53 to 56	Widely scattered severe storms
greater than 56	Scattered severe storms

SWEAT Index (Severe Weather Threat)

The Severe Weather Threat (SWEAT) Index provides a measure of several parameters, including instability, low-level moisture, wind speed, and wind shear. Although its computation (below) is more complicated, the SWEAT Index still can be calculated directly from sounding data. It serves as an important tool for determining both severe thunderstorm potential and tornado potential.

$$SWEAT = 12 \times T_{D\,850} + 20 \times (TT - 49) + 2 \times f_{850} + f_{500} + 125 \times (s + 0.2)\,,$$

where $T_{D\,850}$ = environmental dew point at 850 mb (in °C)
    [Note: If $T_{D\,850} < 0$°C, set $T_{D\,850} = 0$°C],
TT = Total Totals Index [Note: If TT < 49, then set TT = 49],
f_{850} = wind speed at 850 mb (in kts),
f_{500} = wind speed at 500 mb (in kts), and
s = sin(wind direction at 500 mb – wind direction at 850 mb)
    [Notes: Wind directions are in degrees. Set s = 0 if *any* of the following four conditions are false:
    (1) 850-mb wind direction is between 130° and 250°,
    (2) 500-mb wind direction is between 210° and 310°,
    (3) $f_{850} \geq 15$ kts and $f_{500} \geq 15$ kts, or
    (4) wind direction at 500 mb – wind direction at 850 mb > 0].

Table 4 provides an interpretation of the values for the SWEAT Index.

Table 4 – Interpretation of Values for the SWEAT Index

SWEAT Index	Interpretation
greater than 250	Potential for strong convection
greater than 300	Potential for severe thunderstorms
greater than 400	Potential for tornadoes

LABORATORY EXERCISES:

Part I: Adiabatic Processes and the Skew–T Diagram

1. Examine the relationship between the moist and dry adiabats on Figure 3. Explain why the moist adiabats become parallel to the dry adiabats in the upper atmosphere.

_____

_____

_____

_____

2. Examine the horizontal spacing of the saturation mixing ratio lines on Figure 3. On Figure 6, plot saturation mixing ratio as a function of temperature at 1000 mb. In a short paragraph, describe and physically interpret the graph. (Hint: it may be helpful to use a 10-degree temperature interval for your horizontal axis.)

Figure 6 – Graph for Question #2

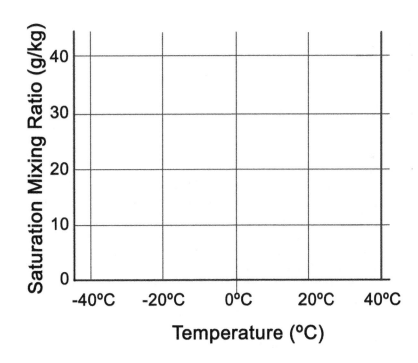

_____

_____

_____

_____

_____

3. For the soundings plotted on Figures 7 and 8, determine the lifting condensation level (LCL), the convective condensation level (CCL), and the equilibrium level (EL).

	ABR (16 April 2002)	**FWD (27 Jan 2000)**
Lifting Condensation Level (LCL)	_____ mb	_____ mb
Convective Condensation Level (CCL)	_____ mb	_____ mb
Equilibrium Level (EL)	_____ mb	_____ mb

Part II: Stability and Stability Indices

4. Using the plotted and tabular data from Aberdeen, SD (ABR) on 16 April 2002 and from Fort Worth, TX (FWD) on 27 January 2000 (Figures 7 and 8; Table 5), calculate the following stability indices.

	ABR (16 April 2002)	**FWD (27 Jan 2000)**
Lifted Index	_____ °C	_____ °C
K–Index	_____	_____
Total Totals Index	_____	_____
SWEAT Index	_____	_____

Figure 7 – Plotted Sounding Data from Aberdeen, SD on 16 April 2002 at 1200 UTC

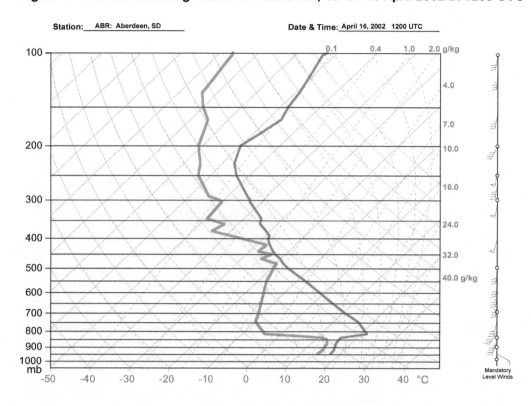

Figure 8 – Plotted Sounding Data from Fort Worth, TX for 27 January 2000 at 1200 UTC

Station: FWD: Fort Worth, TX   Date & Time: January 27, 2000 1200 UTC

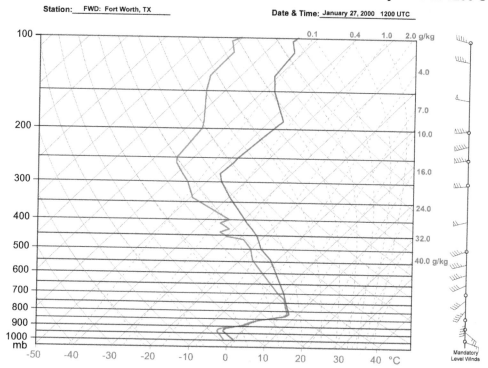

Table 5 – Sounding Data to 500 mb from Aberdeen, SD on 16 April 2002 at 1200 UTC and from Fort Worth, TX for 27 January 2000 at 1200 UTC

72659 ABR 1200 UTC on 16 April 2002						72249 FWD 1200 UTC on 27 Jan 2000					
P	Z	T	T_D	WDIR	WSPD	P	Z	T	T_D	WDIR	WSPD
(mb)	(m)	(°C)	(°C)	(deg)	(kts)	(mb)	(m)	(°C)	(°C)	(deg)	(kts)
1000	20					1000	166				
945	397	18.2	14.9	120	14	996	196	−0.5	−3.0	110	15
925	579	18.2	15.3	150	33	983	304			110	19
913	914			185	43	945	609			120	25
873	1070	16.8	14.6			936	686	−5.3	−6.9		
858	1219			200	43	925	783	−5.5	−7.1	130	25
850	1304	16.6	13.6	205	42	915	868	−5.3	−6.4		
839	1414	16.6	11.8			910	914			140	27
828	1524			215	37	892	1069	−1.3	−1.9		
816	1651	22.2	−3.8			875	1219			165	33
799	1828			210	30	861	1351	0.8	0.4		
771	2133			200	29	850	1456	3.4	3.0	185	45
745	2429	16.6	−9.4			843	1524			185	45
744	2438			195	29	834	1610	7.0	6.5		
718	2743			195	29	820	1749	7.4	6.9		
700	2959	11.2	−10.8	190	31	812	1828			200	47
642	3657			185	34	782	2133			215	43
595	4267			185	33	753	2438			225	41
552	4868	−7.1	−17.1			730	2693	2.2	1.9		
551	4876			185	29	726	2743			230	39
500	5630	−15.1	−18.7	185	32	700	3041	0.4	−1.2	235	37
483	5889	−17.5	−19.2			647	3657			245	39
470	6096			185	35	598	4267			250	43
466	6155	−19.5	−24.5			553	4876			255	43
450	6412	−22.1	−22.9			544	5009	−12.1	−17.0		
441	6560	−23.3	−27.2			500	5660	−17.7	−20.5	250	45

5. Assess and discuss the possibility of thunderstorms and severe thunderstorms based on the indices from the Aberdeen sounding on 16 April 2002 (Figure 7).

_____

_____

_____

_____

6. Assess and discuss the possibility of thunderstorms and severe thunderstorms based on the indices from the Fort Worth sounding on 27 January 2000 (Figure 8).

_____

_____

_____

_____

7. **(Advanced Students/Meteorology Majors)** For 27 January 2000 (Figure 8), the SWEAT Index calculated from the sounding was relatively high. Evaluate and discuss the usefulness of using the SWEAT Index during the winter season.

_____

_____

_____

_____

_____

8. **(Advanced Students/Meteorology Majors)** The SWEAT Index for the 1200 UTC sounding from 3 May 1999 at Norman, OK (OUN) was 295 — less than the threshold for possible tornadoes. A major tornado outbreak, however, occurred across central Oklahoma on that day. Values needed to compute the SWEAT index on 3 May are shown in Table 6. Using the individual terms in the SWEAT Index as your guide, discuss how the structure of the atmosphere may have changed between the 1200 UTC sounding and the time of thunderstorm initiation during mid-afternoon (~3:00 – 4:00 PM CDT).

_____

_____

_____

_____

Table 6 – Values from the 1200 UTC Sounding for 3 May 1999 at Norman, OK (OUN)

Variable	Value
$T_{D\,850}$	8.6° C
TT	48
$WSPD_{850}$	31 kts
$WSPD_{500}$	33 kts
$WDIR_{500}$	250°
$WDIR_{850}$	215°

9. **(Advanced Students/Meteorology Majors)** In light of your findings in question #7, are two launches of radiosondes per day (0000 UTC and 1200 UTC) adequate to assess the stability of the atmosphere? Discuss your conclusion.

_____

_____

_____

_____

_____

_____

10. **(Advanced Students/Meteorology Majors)** If there were funding available for <u>only two soundings per day</u>, are the times when the current radiosondes are launched (0000 UTC and 1200 UTC) the best to assess the stability of the atmosphere for most tornado outbreaks in the continental U.S.? In your discussion, you may want to propose alternative times that would best capture the structure of the pre-convective environment.

_____

_____

_____

_____

_____

_____

Upper-Air Analysis

LAB ACTIVITY OBJECTIVES:

- Using a 500-mb chart, you will practice plotting data measured by the U.S. network of radiosondes.
- Using a 500-mb chart, you will practice and improve your skills contouring meteorological data.
- To better understand the three-dimensional atmosphere, you will practice locating and interpreting particular features on upper-level isobaric charts.
- Given a set of upper-level maps at 850 mb, 700 mb, 500 mb, and 300 mb, you will explain the significant features at each level and interpret some general relationships between features on different maps.

MATERIALS NEEDED:

- Laboratory manual
- Erasable pencil
- Eraser
- Colored pencils
- Calculator (optional)

GLOSSARY:

Dewpoint Depression   *Isotherm*
Geopotential Height   *Jet Stream*
Geostrophic Wind   *Radiosonde*
Hypsometric Equation   *Ridge*
Isobaric Charts   *Trough*

BACKGROUND:

In the United States, **radiosondes** are launched twice daily from approximately 75 stations across the country. These simultaneous releases occur at 0000 and 1200 UTC, as designated by the World Meteorological Organization (WMO). Thus, in the U.S., regular balloon launches occur in the morning and evening, within a few hours of sunrise or sunset. [Some countries, however, only have one launch daily because of the expense of the instrumentation.] The WMO also established mandatory levels (e.g., 850 mb, 700 mb, 500 mb, 300 mb) to ensure that observations around the planet could be assimilated easily and analyzed consistently.

On average, each state has one to two radiosonde sites (see Figure 1). Some states, however, have no sites (e.g., Indiana), while larger states (e.g., Texas) have as many as seven. The stations typically are separated by several hundred miles and are spaced somewhat uniformly.

Figure 1 – Radiosonde Sites in and Around the Contiguous United States

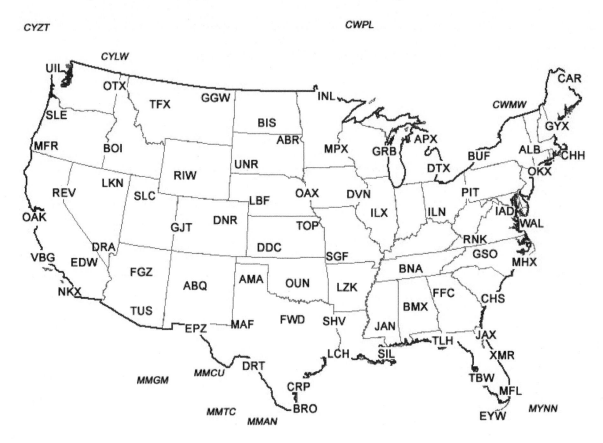

UPPER-AIR STATION PLOTS

Twice daily, maps are plotted for each of the mandatory levels by extracting data for a given level from the available soundings. Because the mandatory levels are defined at pressure levels, the maps sometimes are called constant-pressure charts, or **isobaric charts**. In the U.S., the National Centers for Environmental Prediction (NCEP, an office within the National Weather Service) prepares these upper-air maps for distribution to forecasters and as input to computer forecast models. Aviators also use the upper-air maps to help create flight plans.

Similar to surface plots, upper-air maps include observations displayed in a station model plot (Figure 2). Variables plotted include air temperature (TT), **dewpoint depression** (T-D), **geopotential height** (HGT), height change (H_C; also called height tendency), wind speed, and wind direction. A measure of atmospheric moisture content, dewpoint depression is the difference between air temperature and dewpoint temperature. It is a measure of how close the air is to saturation. Thus, a small dewpoint depression (between 0°C and 3°C) indicates that the air is moist whereas a large dewpoint depression (more than 10°C) indicates that the air is dry.

The geopotential height is calculated from measured data (e.g., temperature, pressure) using the **hypsometric equation**. The plotted heights are in geopotential meters, or gpm, above mean sea level. The height change is the difference between the current geopotential height and that calculated 12 hours earlier. Positive values denote that a given pressure surface has increased in height above the measurement station during the previous 12 hours. The height change is plotted in decameters. Hence, if "–04" were plotted on a 500-mb chart, then the height of the 500-mb pressure surface is 40 meters lower at the current time than 12 hours earlier.

Table 1 displays examples of plotted and decoded radiosonde data from several mandatory levels.

Figure 2 – Standard Plot of Radiosonde Data

TT – Air temperature (in °C; rounded to the nearest °C)
T-D – Dewpoint depression (in °C; rounded to the nearest °C); an "X" is plotted if T-D > 29°C
HGT – Height of pressure surface (in geopotential meters, gpm)
H$_c$ – Height change during the previous 12 hours (in decameters)
Barb – Wind direction (the line from the station dot represents the direction from which the wind is blowing) and wind speed in knots (half line = 5 knots, full line = 10 knots, flag = 50 knots)

Table 1 – Examples of Upper Air Plots

	850 mb	700 mb	500 mb	300 mb	250 mb	200 mb
Wind	Light and Variable	010/20 kts	210/60 kts	270/25 kts	240/30 kts	Missing
TT	22°C	9°C	-19°C	-46°C	-55°C	-60°C
T-D	4°C	17°C	>29°C	not plotted	not plotted	not plotted
Dew Point	18°C	-8°C	Dry	Dry	Dry	Dry
HGT	1,479 m	3,129 m	5,580 m	9,190 m	10,370 m	11,910 m
H$_c$	not plotted	- 30 m	+ 30 m	+ 100 m	+ 10 m	not plotted

As evident by the examples in Table 1, decoding the height value on the upper-air plots varies depending on the pressure level. Table 2 indicates how to decode the height of an upper-air observation on a given isobaric map.

Table 2 – Decoded Upper–Air Height Values

Pressure Level	Approx. Height	Coded Height	Actual Height	Additional Information
850 mb	1500 gpm	479	1479 gpm	Add 1000 gpm.
700 mb	3000 gpm	129	3129 gpm	If > 500 gpm, add 2000 gpm. Otherwise, add 3000 gpm.
500 mb	5500 gpm	558	5580 gpm	Multiply by 10.
300 mb	9000 gpm	919	9190 gpm	If > 500, multiply by 10. Otherwise, multiply by 10 and add 10,000.

LABORATORY EXERCISES:

Part I: 500–mb Analysis

1. Using an erasable pencil, plot the following 500–mb observations on Figure 3, according to the plotting model in Figure 2. Use Figure 1 for the radiosonde locations. (Note: Table 3 contains measurements expressed to tenths of degrees and tenths of knots. You may need to round the data to plot according to the standard conventions.)

Table 3 – Selected 500–mb Observations from 0000 UTC on 17 October 1996

Station ID	Station Name	Latitude	Longitude	Height (in m)	Temp (in °C)	Dewpoint (in °C)	Wind Speed (in kts)	Wind Direction (in deg)
ABR	Aberdeen, SD	45.45° N	98.43° W	5590	−16.3	−19.2	48.6	235
AMA	Amarillo, TX	35.23° N	101.70° W	5740	−11.1	−36.1	44.7	285
BNA	Nashville, TN	36.13° N	86.68° W	5770	−12.5	−22.5	15.5	260
EYW	Key West, FL	24.55° N	81.75° W	5880	−4.9	−13.9	7.8	95
GGW	Glasgow, MT	48.22° N	106.62° W	5470	−30.5	−45.5	9.7	260
GJT	Grand Junction, CO	39.12° N	108.53° W	5620	−23.1	−24.6	48.6	255
OTX	Spokane, WA	47.68° N	117.63° W	5500	−31.7	−38.7	25.3	305
OUN	Norman, OK	35.22° N	97.45° W	5750	−10.9	−31.9	29.1	245
TFX	Great Falls, MT	47.45° N	111.38° W	5460	−30.9	−43.9	19.4	230
TUS	Tucson, AZ	32.12° N	110.93° W	5800	−9.9	−45.9	17.5	325

2. Analyze the 500-mb height field on Figure 3 by drawing height contours at intervals of 60 m (6 decameters, or 6 dam). Use an erasable pencil, and use the accepted contour values of 5340, 5400, 5460... meters (534, 540, 546 dam). You may find it useful to start with the height contour of 552 dam (in Central WY). When you are satisfied with the placement of your height lines, trace over them with a black colored pencil to solidify your analysis. [Note: Because the observations are rounded to the nearest 10 m, a plotted value of 564 dam could represent a 500-mb height from 5635 m to 5644 m. This degree of uncertainty in the plotted measurements allows some freedom in drawing and smoothing the height contours. The overall smoothness of the contour is more important than "forcing a fit" of a given contour line to any one observation.]

3. Using a solid, red zigzag line, mark the height **ridge**(s) on the 500-mb map in Figure 3. Similarly, use a dashed, red straight line to indicate any height **trough**(s).

4. On your 500-mb analysis (Figure 3), (a) where are the strongest **geostrophic winds** and (b) where are the weakest geostrophic winds? (c) Discuss what you see as the relationship between the strength of the geostrophic winds and the spacing (distance) between the height contours.

Location(s) of strongest geostrophic winds _____

Location(s) of weakest geostrophic winds _____

_____

_____

_____

Figure 3 – 500–mb Chart for 0000 UTC on 17 October 1996

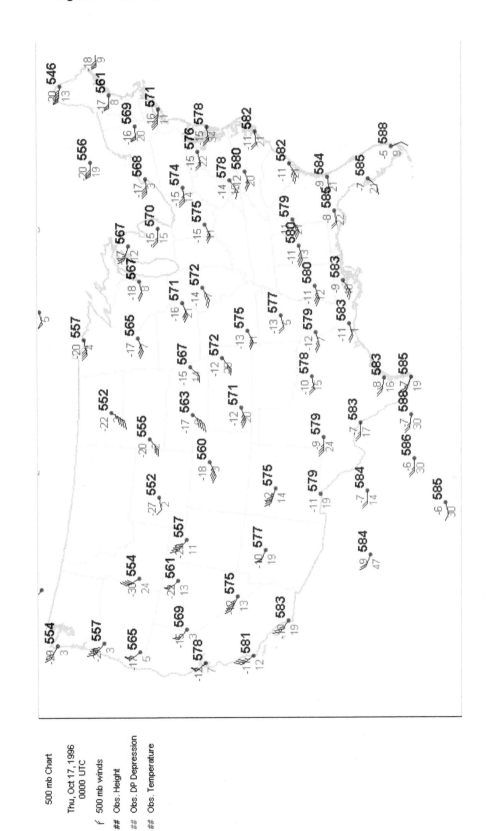

500 mb Chart

Thu, Oct 17, 1996
0000 UTC

ʄ 500 mb winds

Obs. Height

Obs. DP Depression

Obs. Temperature

5. In general, (a) do the 500-mb heights increase or decrease from north to south over the United States (e.g., from Minnesota to Texas)? (b) Explain why they increase or decrease.

_____

_____

_____

_____

6. Comment on the temperature pattern as it relates to the orientation and alignment of the troughs and ridges.

_____

_____

_____

_____

7. *(Advanced Students/Meteorology Majors)* Explain why 500 mb was chosen as a mandatory level and is routinely analyzed. [Hint: Think about the resulting motion of a 30-m tall pole from horizontal forces applied equally at heights of 5 m, 15 m, and 25 m.]

_____

_____

_____

_____

Part II: 850–mb, 700–mb, and 300–mb Analyses

In addition to the 500-mb map, meteorologists examine surface, 850-mb, 700-mb, and 300-mb (or 250-mb) charts to understand the current state of the atmosphere in three dimensions. In this section, you will work to understand the three-dimensional picture for 17 October 1996. Figures 4, 5, and 6 display the isobaric maps for 850 mb, 700 mb, and 300 mb, respectively. Figure 7 displays the surface map. Use these figures to answer the following questions.

8. Calculate the dewpoint temperature at 850 mb (Figure 4) for the following stations: Boise, ID; Bismarck, ND; Buffalo, NY; Key West, FL; Nashville, TN; Moorehead City, NC; Norman, OK; Oakland, CA; Omaha, NE; and Topeka, KS. Write the dew point value in green on Figure 4 above the corresponding station temperature.

9. The 850-mb chart is used frequently to analyze the amount of moisture available for clouds and precipitation. Using Figure 4, where is the driest air in the United States at 850 mb? Where is the maximum moisture at 850 mb?

Location of driest air at 850 mb _____

Location of maximum moisture at 850 mb _____

Figure 4 – 850–mb Chart for 0000 UTC on 17 October 1996 with Height Contours (thick, gray lines) and Isotherms (thin, red lines)

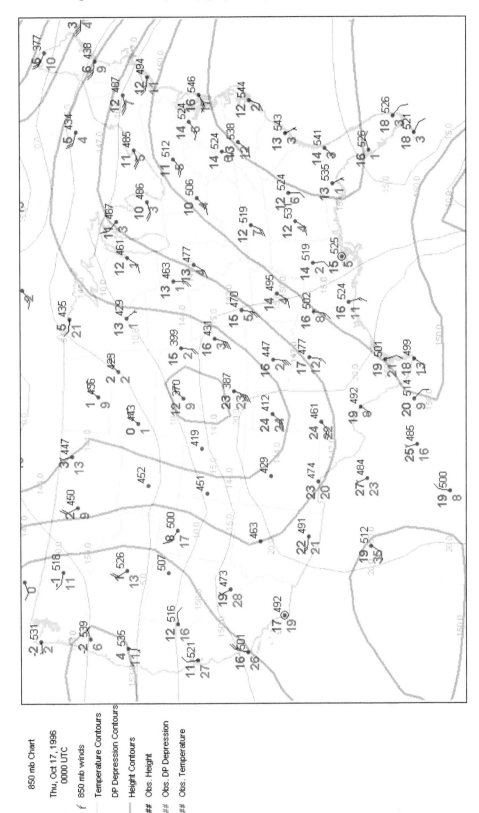

Figure 5 – 700–mb Chart for 0000 UTC on 17 October 1996
with Height Contours (thick, gray lines) and Isotherms (thin, red lines)

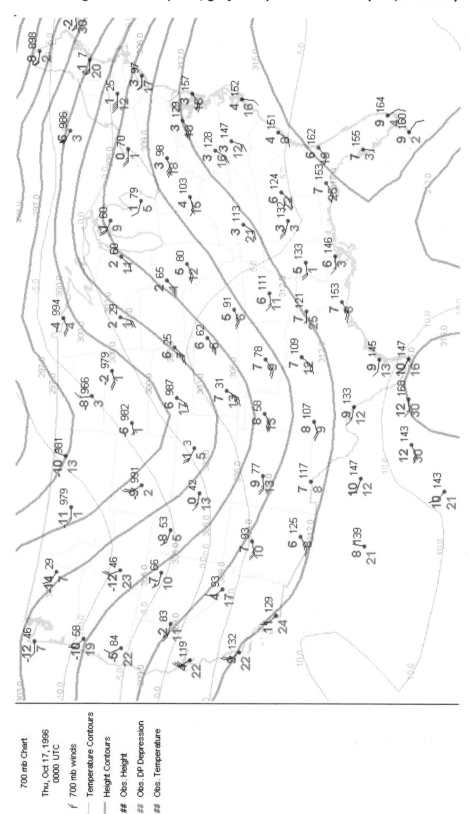

700 mb Chart

Thu, Oct 17, 1996
0000 UTC

⌐ 700 mb winds

— Temperature Contours

— Height Contours

Obs. Height

Obs. DP Depression

Obs. Temperature

Figure 6 – 300–mb Chart for 0000 UTC on 17 October 1996
with Height Contours (thick, gray lines)

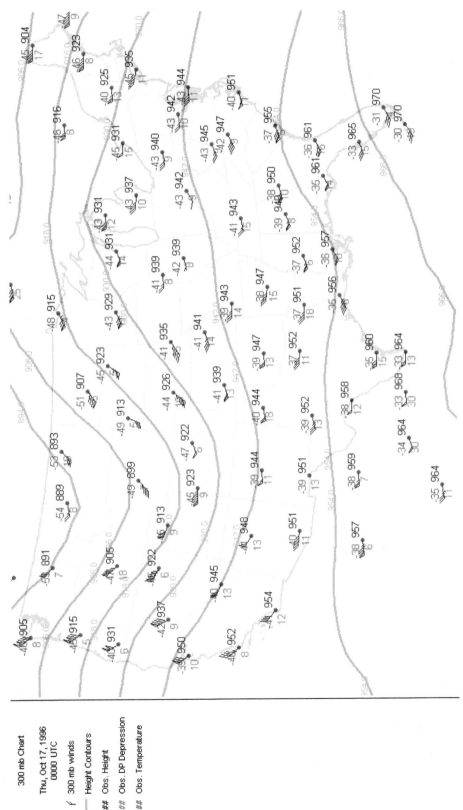

300 mb Chart

Thu, Oct 17, 1996
0000 UTC

🏳 300 mb winds

――― Height Contours
Obs. Height
Obs. DP Depression
Obs. Temperature

10. Use the 700-mb chart (Figure 5) and a green pencil to outline and shade the areas where the dewpoint depression (T-D) is less than or equal to 3°C. These shaded areas locate regions where there is substantial moisture and perhaps extensive cloud cover. In addition to the 850-mb chart, the 700-mb chart is used to analyze the amount of moisture available for clouds and precipitation.

11. If you were a pilot and your license limited you to "visual flight rule" conditions (i.e., clear skies), what area(s) of the country would you most likely remain grounded?

_____

_____

12. Using Figure 5, what is the direction of the geostrophic wind at 700 mb over the northwest U.S.? Based only on this geostrophic wind and the configuration of the **isotherms** (Figure 5), would you expect the 700-mb temperature over Colorado to increase, decrease, or stay the same during the next 12 hours?

Direction of geostrophic wind at 700 mb over NW U.S. _____

Temperature over CO will _____ during next 12 hours

13. The 300-mb chart is used frequently to analyze the location and strength of the **jet stream** (i.e., a relatively narrow current of strong winds) and its influence on weather systems. The jet stream occurs near this level because the geostrophic winds are faster here than at the other mandatory levels. On the 300-mb chart (Figure 6), draw a thick, blue curved line where you think the jet stream is located. The jet stream should be parallel to the flow and extend from one side of the map to the other. [Hint: Start by finding the fastest wind speeds.]

14. Compare the isobaric charts for 700 mb (Figure 5), 500 mb (Figure 3), and 300 mb (Figure 6). Discuss how the trough over the west-central U.S. changes its location with height.

_____

_____

_____

_____

15. Compare the 500-mb chart (Figure 3) with its corresponding surface map (Figure 7). Geographically, where are the 500-mb troughs/ridges located as compared to the surface highs and lows?

_____

_____

_____

_____

Figure 7 – Surface Chart for 0000 UTC on 17 October 1996 with Isobars (solid, blue lines)

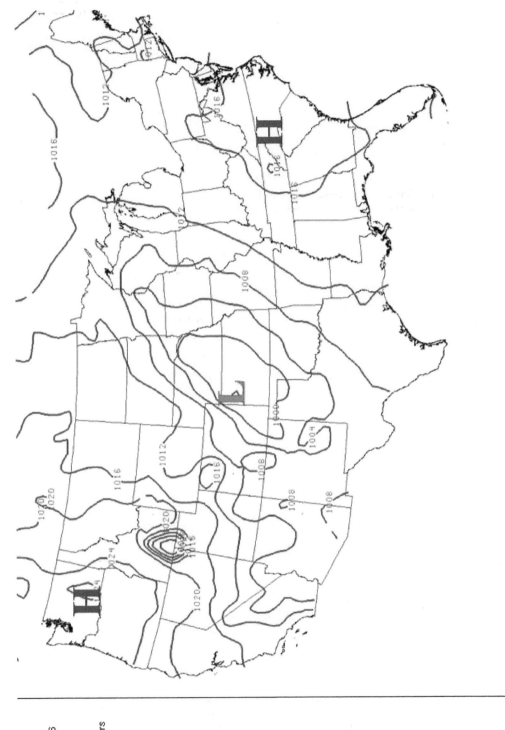

Surface Chart

Thu, Oct 17, 1996
0000 UTC

Pressure Contours

Thunderstorms and Their Environment

LAB ACTIVITY OBJECTIVES:

- You will explain the processes involved in creating, sustaining, and ending thunderstorm activity.
- You will discriminate between different types of thunderstorms and describe how these differences influence storm development.
- Given appropriate maps of atmospheric measurements (e.g., temperature, dew point, winds, radar reflectivity), you will associate these measurements with the initiation, growth, and decay of thunderstorms.
- Given appropriate maps of atmospheric measurements (e.g., temperature, dew point, winds, radar reflectivity), you will locate phenomena associated with thunderstorm activity, including fronts, drylines, thunderstorm outflows, and, for the advanced student, heatbursts.
- The advanced student will explain how and why the dryline moves eastward during the day and westward during the night.
- The advanced student will describe the conditions favorable for the development of heatbursts.

MATERIALS NEEDED:

- Laboratory manual
- Pencil or pen
- Colored pencils

GLOSSARY:

Cloud-to-Ground Lightning
Convective Precipitation
Convergence
Cumulus Stage (of a Thunderstorm)
Developing Stage (of a Thunderstorm)
Dissipating Stage (of a Thunderstorm)
Downburst
Downdraft
Dryline
Gust Front
Heatburst

Lifting Mechanism
Mature Stage
  (of a Thunderstorm)
Microburst
Multi-Cell Thunderstorm
NEXRAD
Orography
Outflow
Radar
Severe Thunderstorm
Single-Cell Thunderstorm

Squall Line
Straight-Line Winds
Stratiform Precipitation
Supercell Thunderstorm
Thunderstorm Outflow
Tornado
Towering Cumulus
Unstable Air
Updraft
Wind Shear

BACKGROUND:

Thunderstorms provide the majority of warm season precipitation in the United States. Their heavy rains replenish the water supply to lakes, streams, and aquifers; their lightning strikes provide vital nitrogen compounds to the soil, thereby acting as a natural fertilizer; and their winds act to remove weakening trees from the forest, thus stimulating renewal of these natural resources. From a meteorological perspective, thunderstorms also play a critical role in the planetary energy budget by transporting heat energy away from the surface and into the upper atmosphere.

In the United States, **severe thunderstorms** are defined as thunderstorms that contain winds greater than 50 knots (58 mph), hail greater than 3/4 inch, or a **tornado**. Although the public often does not react as quickly to severe thunderstorm warnings as to tornado warnings, severe thunderstorms can be as deadly and as damaging to crops and property. **Cloud-to-ground lightning** kills more people than tornadoes, and any thunderstorm can contain cloud-to-ground lightning. Tornadoes occur infrequently, but can contain winds that exceed 200 mph. The greatest threat to life and property in a tornado is wind-blown debris, including vehicles that can be lifted off the roadway.

LIFE CYCLE OF A THUNDERSTORM

Thunderstorms have a three-stage life cycle, as depicted by Figure 1. The **cumulus**, or **developing**, **stage** is characterized by a **towering cumulus** cloud composed mainly of an ascending column of air called the **updraft**. In this stage, which may last for 10-15 minutes, little or no rain falls. The **mature stage** occurs when the storm reaches its maximum height (e.g., 40,000 feet or higher). The mature storm can produce heavy, **convective precipitation**, hail, frequent lightning, and strong winds. The **dissipating stage** occurs when the storm is dominated by **downdrafts**, cutting off the supply of warm, moist, **unstable air** that feeds the storm's updraft. **Straight-line winds** or a **downburst** may accompany the dissipating stage. In addition, the height of the storm and the rate of rainfall decrease when the storm collapses. Lightning is a danger during *any* stage of a thunderstorm.

Lightning can strike several miles away from the core of a thunderstorm without warning. Lightning safety rules for outdoor activities take into account the potential for sudden lightning events. The "30–30 rule" states that outdoor activities should be suspended if fewer than 30 seconds elapse between the sight of lightning and the sound of thunder. The activity may be resumed safely 30 minutes after the last flash of lightning or clap of thunder.

THUNDERSTORM CLASSIFICATION

Thunderstorms are classified into four types: single-cell, multi-cell, supercell, and squall line. **Single-cell thunderstorms** are short-lived (~20–30 minutes) and cover geographic areas of only a few square miles. Single-cell thunderstorms develop in environments with little vertical **wind shear**. As a result, these storms form upright and their rain falls directly though the updraft column, suppressing the upward motion required to maintain the storm.

Multi-cell storms are formed by two or more single-cell storms located near one another. As a result, complex interactions can occur between the updraft of one storm and the downdrafts of neighboring storms. New storms can develop toward the rear of the complex every 5 to 15 minutes.

Although single-cell and multi-cell storms can produce severe weather, **supercell storms** almost always produce severe weather. These prolific severe storms often persist for several hours and travel several hundred miles. The supercell develops in an environment with strong vertical wind shear (e.g., stronger winds aloft than at the surface). As a result, the updrafts of supercells are tilted with height, physically separating the updraft and downdraft regions (Figure 2). The separation of the updraft and downdraft allows precipitation to fall without destroying the storm or its updraft. Most supercells also rotate as a result of the vertical wind shear being deflected into the horizontal. Most violent and damaging tornadoes are associated with supercells.

A **squall line** is a line of active thunderstorms that may extend across several hundreds of miles. Any combination of single-cell, multi-cell, or supercell storms forms the squall line. Squall lines often are associated with cold fronts and **drylines**. A large area of **stratiform precipitation** sometimes follows a squall line. Weak tornadoes may occur at the leading edge of a squall line, and strong tornadoes may accompany any supercell that is embedded in the squall line.

THE ENVIRONMENT OF THUNDERSTORMS

Thunderstorms require a supply of unstable air (usually warm and moist) for fuel. They also require a source of lift, allowing deep clouds to initiate in the unstable air. The **lifting mechanism** can originate at the surface (e.g., from **orography**, a front, or a dryline) or aloft (e.g., from an upper-level storm system or the jet stream).

Gust Fronts

In their wake, thunderstorms normally leave a footprint of more stable air (usually cooler and drier) from the downdraft. The boundary between the warm, humid environment and the cool, dry **thunderstorm outflow** is called a **gust front** (see Figure 3). As the gust front moves away from its parent thunderstorm, it can act as a lifting mechanism to help generate additional storms.

Figure 1 – The Life Cycle of a Thunderstorm

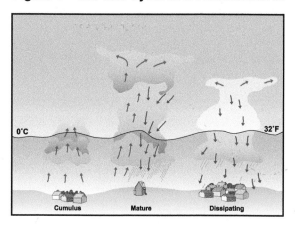

Drylines

Although drylines occur frequently across the Southern Plains of the United States, they are rare worldwide. In the U.S. Southern Plains, the dryline becomes established when southerly or easterly winds advect moisture from the Gulf of Mexico and westerly winds advect dry air from the elevated terrain of the Rockies and Mexican Plateau. These air masses with vastly different moisture characteristics most frequently meet across eastern New Mexico, eastern Colorado, western Texas, western Oklahoma, and western Kansas. The boundary between the continental tropical and maritime tropical air masses is the dryline.

Drylines are characterized by a significant moisture gradient. As a dryline passes a given location, a large, abrupt decrease in dewpoint temperature will be observed. Like a front, the dryline generates **convergence** and, thus, rising motion, becoming a focus for thunderstorm development. Figures 4 and 5 depict the structure and daily evolution of a Southern Plains dryline.

Downbursts

During their decaying stages, thunderstorms can produce sudden, localized bursts of wind from their downdrafts. These downbursts can produce significant damage to property and vegetation. Because they typically extend across a few miles for 15 to 20 minutes, downbursts are difficult to observe routinely and are even harder to predict. Because much remains to be understood about downbursts, research scientists continue to study the mechanisms that initiate downbursts within thunderstorms. Two types of downbursts are **microbursts** and **heatbursts**.

Microbursts

Microbursts are so named because of their small horizontal extent (less than 4 km across). Microbursts can form when a moist downdraft encounters dry air beneath the cloud base of a thunderstorm. Evaporation, and its associated cooling, occurs as the downdraft moves through the dry layer. This cool, dense air sinks and enhances the original downdraft. When the downdraft hits the ground, the cool air diverges (i.e., spreads out) in all directions. Because of wind shear generated at the ground, microbursts are extremely dangerous for aircraft attempting to take off or land.

Figure 2 – Schematic of a Supercell Thunderstorm

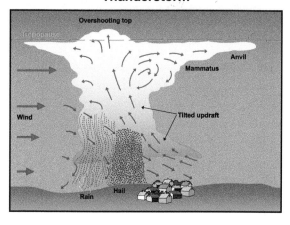

Figure 3 – Schematic of the Interaction of a Thunderstorm with Its Environment

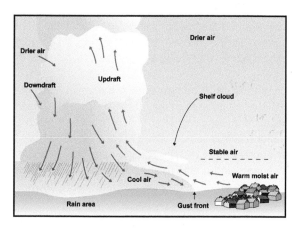

Figure 4 – Schematic of a Southern Plains Dryline during Early Morning

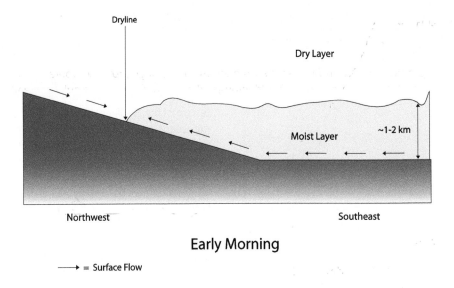

Dryline

Dry Layer

Moist Layer

~1-2 km

Northwest

Southeast

Early Morning

⟶ = Surface Flow

Figure 5 – Schematic of a Southern Plains Dryline during Mid–Afternoon

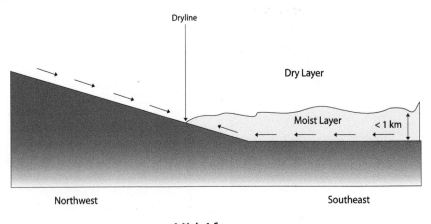

Dryline

Dry Layer

Moist Layer

< 1 km

Northwest

Southeast

Mid-Afternoon

⟶ = Surface Flow

Heatbursts

Heatbursts are small-scale, warm anomalies characterized by a large, abrupt increase in temperature and decrease in dewpoint temperature beneath a collapsing thunderstorm. Heatbursts form when rain within a thunderstorm downdraft completely evaporates before the descending air arrives at the ground. Because descending unsaturated air compresses and warms dry adiabatically, the downdraft warms significantly after the rain evaporates. As a result, the warm (or hot), dry downdraft strikes the ground and spreads outward, often accompanied by strong winds. Heatbursts are often warm-season events and occur during the night. Surface temperatures measured during a heatburst can be 90° to 100°F well after sunset.

Once thought to be rare events, the small horizontal scale of heatbursts makes them difficult to observe because heatbursts often occur between weather stations. A mesoscale network, however, can observe heatbursts more readily. For example, the Oklahoma Mesonet (with a station spacing of less than 30 km) has documented more than 100 heatbursts between 1994 and 2002.

Severe Thunderstorms: A Case Study

LEARNING OBJECTIVES:

- Using the forecast funnel technique and your knowledge of thunderstorm development, you will develop a forecast of the storm initiation and storm motion for a severe thunderstorm event.
- You will combine the important attributes of several maps and datasets into a synoptic-scale composite map and a mesoscale composite map.
- Using key figures, you will describe relationships between data from a NEXRAD radar site and from a network of surface observing stations.

MATERIALS NEEDED:

- Laboratory manual
- Pencil or pen
- Colored pencils
- Radar animation supplied by instructor

GLOSSARY:

Base Reflectivity
Clear-Air Mode
Composite Chart
Forcing
Forecast Funnel
Instability
Mesoscale
Moisture Tongue
NEXRAD

Precipitation Mode
Ridge
Splitting Storm
Storm Scale
Synoptic Scale
Thermal Ridge
Thin Line
Trough

BACKGROUND:

Atmospheric motions occur on a variety of horizontal space scales, ranging from planetary motions to turbulent motions on the molecular scale. Complex interactions occur between the scales. Some of these interactions are not well understood and, hence, are the subject of much research. In weather forecasting, important elements to understand include forces on the **synoptic scale** (i.e., horizontal size on the order of 1000 km and a time scale of several days), **mesoscale** (i.e., tens to hundreds of kilometers and a single day), and **storm scale** (i.e., few kilometers and from minutes to hours). One technique that forecasters use to mentally assimilate information across many spatial and temporal scales is the **forecast funnel**. This technique begins with the analyses of synoptic-scale features followed by successive analyses of smaller scale features.

Because meteorologists use charts at multiple levels in the atmosphere (e.g., 300 mb, 500 mb, surface), satellite images, radar data, and other data, the process to synthesize these varied types of information can be strenuous. One tool sometimes used to help the process is the **composite chart** — a chart constructed from information from multiple sources. A composite chart may indicate, for example, the relative positions of the 500-mb **trough**, the jet stream at 300 mb, surface characteristics, and cloud cover. In this laboratory exercise, the forecast funnel and composite chart will be applied to a severe weather event on 2 April 1994.

SEVERE WEATHER OF 2 APRIL 1994

Parts of Oklahoma experienced a severe weather episode on 2 April 1994. This event embodied a classic severe weather event, exhibiting many of its required ingredients. For example, surface features included both a cold front and a dry line. An upper-level trough provided synoptic-scale **forcing**. Steep lapse rates (i.e., very cold air lying above much warmer air) and moderately high dew points (especially for early April) acted to create **instability**. Storm-scale features included **splitting storms**. Although this case did not produce a widespread severe weather outbreak, it did produce large hail, severe winds, and a couple of tornadoes. Even so, these localized events can cause significant damage. In this case, Pontotoc County in Oklahoma received hail damage of more than $3 million, including $100,000 at a car dealership and $300,000 at six school buildings.

LABORATORY EXERCISES:

In this laboratory exercise, the forecast funnel technique will be applied by creating and by analyzing two composite charts. The composite charts will be constructed using the symbols in Figure 1. The synoptic-scale composite chart will display features from several upper-air maps of the United States. Synoptic-scale composite charts typically highlight areas which span multiple states and provide an overview of the ingredients for a given weather situation (e.g., large-scale rising motion, availability of moisture).

A mesoscale composite chart also will be constructed, using radar data from NWS's NEXRAD and surface observations from the Oklahoma Mesonet. The mesoscale chart will narrow the forecaster's focus to a smaller area (e.g., a few counties) that highlight important features that might not be detected using only the synoptic-scale chart (e.g., the locations of a cold front or a dry line).

Figure 1 – Several Symbols Used on a Composite Chart

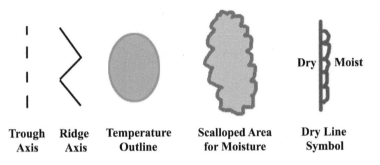

| Trough Axis | Ridge Axis | Temperature Outline | Scalloped Area for Moisture | Dry Line Symbol |

Part I: Synoptic–Scale Composite Map

1. Using Figure 2, locate and highlight, with a black dashed line, the 500-mb trough axis. Also, locate and highlight, with a black zigzag line, the 500-mb **ridge** axis located downstream from the trough. Transpose both of these axes to Figure 3 — your synoptic-scale composite chart.

Figure 2 – 500–mb Chart for 1200 UTC on 2 April 1994

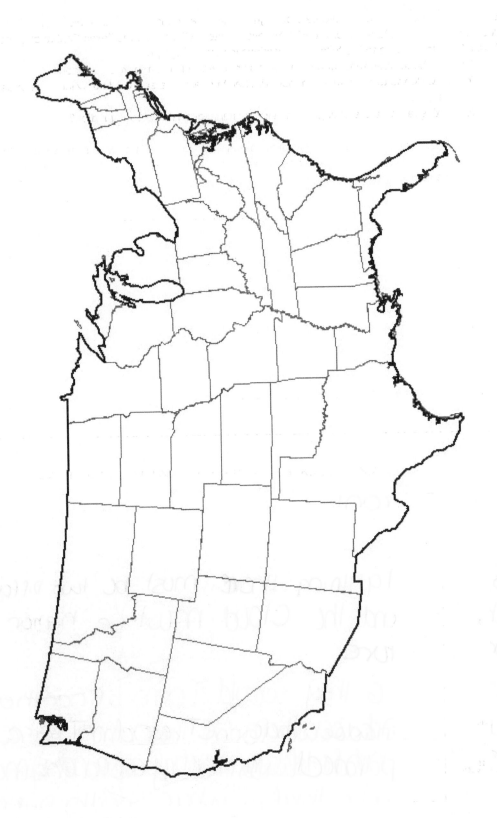

2. The area between the axes of an upper-level trough and the next downstream ridge often is a preferred region for large-scale rising motion. Outline this area on your composite chart (Figure 3) using a blue pencil.

3. Examine the 850-mb temperature field on Figure 4 and the sounding data in Table 1. In a short paragraph, describe the temperature pattern at 850 mb over western Oklahoma, the Texas Panhandle, and southwest Texas. Include answers to the following questions in your paragraph:
    (a) Are the temperatures at 850 mb in this area relatively warm or relatively cool?
    (b) How do the temperatures change with height over this area (i.e., do they increase or decrease with height)?
    (c) Is this temperature change with height a stable or unstable configuration?
    (d) Based upon the wind and moisture data (Table 1), from where did the air at 850 mb in this region likely come? (Hint: You may want to compare the 850-mb temperatures and winds with the surface temperatures and winds. You also may want to calculate the temperature change with height from the sounding data in Table 1.)

_____

_____

_____

_____

_____

_____

_____

Figure 4 – 850–mb Temperature Chart for 1200 UTC on 2 April 1994

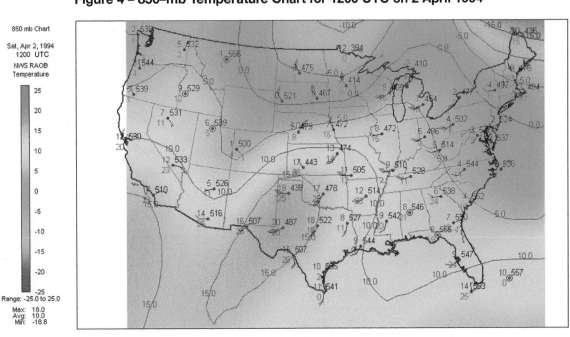

**Table 1 – Selected Mandatory–Level Data for Oklahoma and Texas
for 1200 UTC on 2 April 1994**

Station	Surface					925 mb				850 mb			
	T (°C)	T_D (°C)	P_{STN} (mb)	Wind Dir (deg)	Wind Spd (kts)	T (°C)	T_D (°C)	Wind Dir (deg)	Wind Spd (kts)	T (°C)	T_D (°C)	Wind Dir (deg)	Wind Spd (kts)
AMA*	9.4	5.2	886**	210	14	N/A	N/A	N/A	N/A	18.0	–7.0	265	35
MAF*	11.8	9.2	914**	170	13	N/A	N/A	N/A	N/A	19.6	–0.4	235	32
OUN*	13.0	8.9	971**	180	12	13.8	9.7	210	45	16.6	–8.4	210	35
SEP*	10.8	7.1	971**	180	8	10.6	8.9	205	35	17.6	–0.4	205	24

* AMA = Amarillo, TX; MAF = Midland, TX; OUN = Norman, OK; SEP = Stephenville, TX (just southwest of Fort Worth)

** P_{STN} is the raw station pressure observed by the barometer and is not reduced to sea level.

4. On your composite chart (Figure 3), outline in red the significant temperature pattern discussed in question 3. (Hint: You may want to sketch a 15°C or 16°C isotherm to represent *the pattern* in this case.)

5. Examine the 850-mb dewpoint depression maps (Figures 5 and 6). Compare the moisture pattern at 1200 UTC on 2 April 1994 (Figure 5) with the one 12 hours later at 0000 UTC on 3 April (Figure 6). What happened to the moisture over Oklahoma and Texas? Scallop in green the location of the significant moisture at 850 mb at 1200 UTC on your composite chart (Figure 3). Use a green arrow to denote the movement of the moistest air (also known as a **moisture tongue**) during the following 12-hour period (to 0000 UTC on 3 April).

Figure 5 – 850–mb Dewpoint Depression Map for 1200 UTC on 2 April 1994

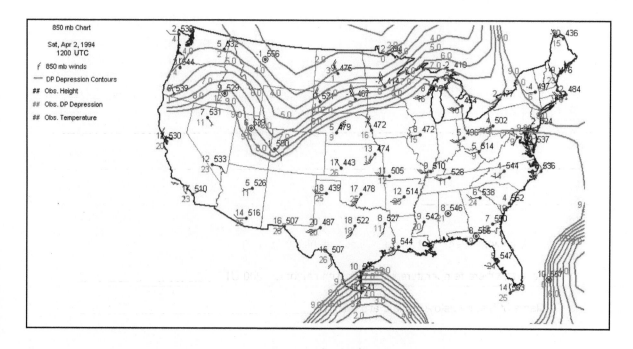

Figure 6 – 850–mb Dewpoint Depression Map for 0000 UTC on 3 April 1994

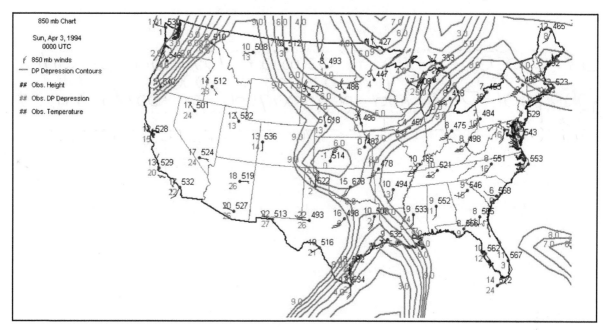

6. Examine the Skew-T diagrams for Norman, OK (OUN) at 1200 UTC on 2 April 1994 (Figure 7) and 0000 UTC on 3 April 1994 (Figure 8). On the morning sounding, describe the structure of the temperature plot beneath 500 mb. List the levels where the temperature decreases with height and the layers where temperature increases with height. What is the technical name for the significant feature or pattern below 850 mb?

_____

_____

_____

_____

_____

_____

_____

Layer(s) where temperature decreases with height at 1200 UTC _____

Layer(s) where temperature increases with height at 1200 UTC _____

Name of feature below 850 mb at 1200 UTC _____

**Figure 7 – Skew T–Log P Plot of Sounding Data from Norman, OK (OUN)
for 1200 UTC on 2 April 1994**

72357 OUN Norman, OK

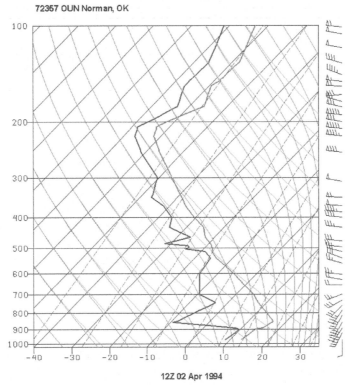

12Z 02 Apr 1994

**Figure 8 – Skew T–Log P Plot of Sounding Data from Norman, OK (OUN)
at 0000 UTC on 3 April 1994**

72357 OUN Norman, OK

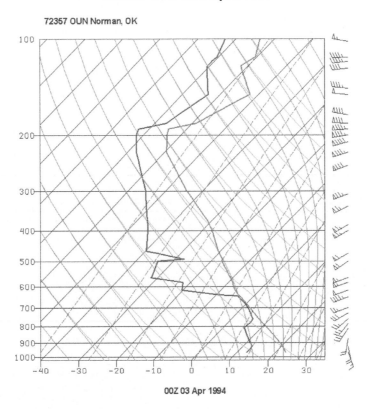

00Z 03 Apr 1994

7. **(Advanced Students/Meteorology Majors)** Does the temperature pattern below 850 mb aid or oppose storm development? Explain your answer.

_____

_____

_____

_____

_____

_____

8. Look at the winds on the morning sounding (Figure 7). Comment on how the winds speed and direction change with height. Often, meteorologists consider 500-mb winds to be the steering currents for storms. Based on this concept, what direction should the storms move across central Oklahoma on the afternoon on 2 April 1994? To indicate your forecasted storm motion, plot and label a blue arrow on your composite map across central Oklahoma.

Forecasted direction of storm movement across central OK _____

9. Compare the morning sounding (Figure 7) to the evening sounding (Figure 8). Describe how the winds changed, in general, throughout the day.

_____

_____

_____

_____

_____

_____

Part II: Mesoscale Composite Map

County-level data from the Oklahoma Mesonet (www.mesonet.org) was available for the mesoscale analysis on 2 April 1994. Mesonet maps are displayed in local time. You may want to refer to the map of Oklahoma counties in Figure 9 to use county names in your answers to the questions that follow.

10. Examine the temperature, dew point, and wind data for 2:30 PM CST on 2 April 1994 (Figures 10, 11, and 12). Identify any fronts, drylines, and high- or low-pressure centers using the customary symbols and colors. Transpose those symbols onto your mesoscale composite map (Figure 13).

11. Using Figure 10, identify the location of the warmest air (also known as a **thermal ridge**). Shade this warm air region in red pencil on your mesoscale composite map (Figure 13). [Hint: To best identify this feature, you may want to locate and sketch an isotherm in the 80° – 83°F range.]

12. Using Figure 11, identify the location of the moisture tongue. Draw a green scalloped line around this moist air region on your mesoscale composite map (Figure 13). Shade the area within the scalloped line with a green pencil. [Hint: To best identify this feature, you may want to locate and sketch the 55°F isodrosotherm. Surface dewpoint temperatures in the mid-50s often are considered as the threshold needed for severe storm development.]

13. Inspect the **base reflectivity** image from 2:42 PM CST (2042 UTC) shown by Figure 14. At this time, the **NEXRAD** radar was operating in **clear-air mode** — a very sensitive configuration used to detect atmospheric boundaries such as fronts or drylines. In clear-air mode, the radar detects changes in atmospheric density, typically resulting from either temperature or moisture gradients. When two dissimilar air masses are adjacent, a discontinuity in atmospheric density may show up as a **thin line** on the radar display.

    On Figure 14, a thin line is evident to the west of the radar's location (near Oklahoma City). How does the thin line correspond with patterns observed by the Oklahoma Mesonet (Figures 10, 11, and 12)?

    _____

    _____

    _____

    Describe the density differences between the west and east sides of the thin line (i.e., Is the atmospheric density higher or lower on the west side as compared to the east?). How did you arrive at your answer?

    _____

    _____

    _____

    Explain why weather radar may have an advantage over a network of surface weather stations in detecting these low-level discontinuities.

    _____

    _____

    _____

Figure 9 – Map of the Names of Oklahoma Counties

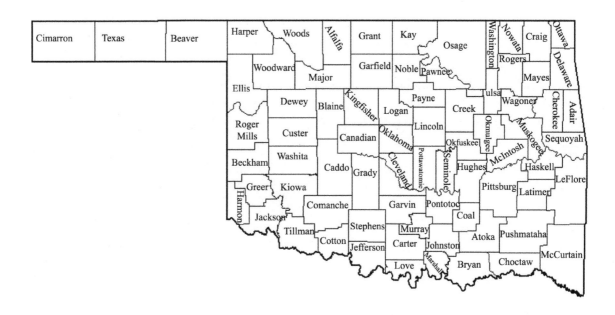

Figure 10 – Temperature (in °F) for 2 April 1994 at 2:30 PM CST (2030 UTC), as Measured by the Oklahoma Mesonet

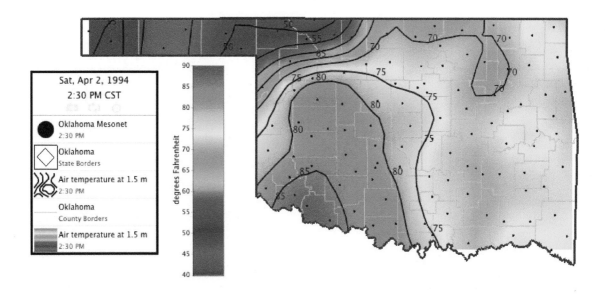

Figure 11 – Dew Point (in °F) for 2 April 1994 at 2:30 PM CST (2030 UTC), as Measured by the Oklahoma Mesonet

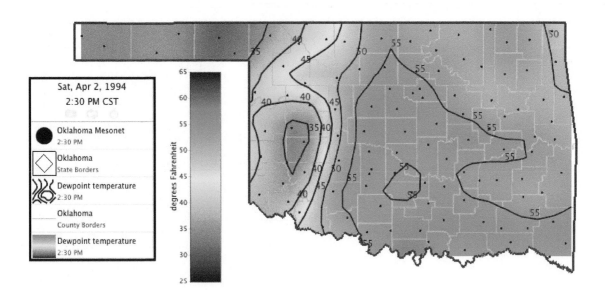

Figure 12 – Winds Speed (in kts) and Direction for 2 April 1994 at 2:30 PM CST (2030 UTC), as Measured by the Oklahoma Mesonet

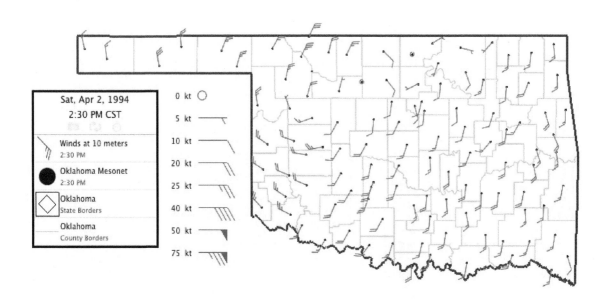

Figure 13 – Blank Map for Mesoscale Composite Chart Valid at 2:30 PM (2030 UTC) on 2 April 1994

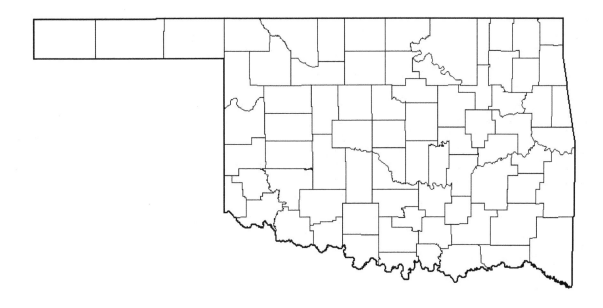

Figure 14 – Base Reflectivity (at 0.5° tilt) in Clear Air Mode for 2:42 PM CST (2042 UTC) on 2 April 1994, as Measured by the NEXRAD Radar near Oklahoma City, OK

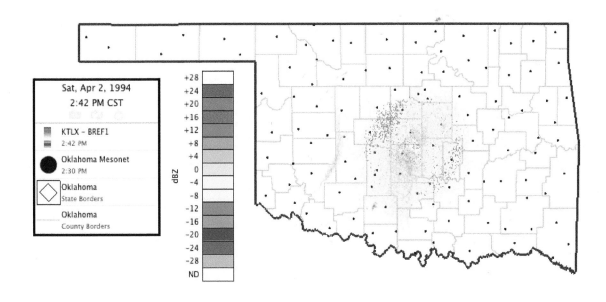

14. By 2:30 PM CST, thunderstorms had not developed in Oklahoma. Using both your knowledge of the environment represented by your composite charts (Figures 3 and 13) and your knowledge of the ingredients necessary for thunderstorm development, where will thunderstorms most likely develop first? Outline your forecasted area of several counties (~5-10) in black on your mesoscale composite chart (Figure 13) where you think the storms will form.

Counties where storms will form initially (Note: County names are on Figure 9.) _____

_____

15. When you finish your lab, your instructor will show you a radar loop of the base reflectivity (in **precipitation mode**) from 3:15 PM to 8:00 PM CST. Use this information to verify qualitatively your forecasts of storm development and motion. Describe how well you forecasted the storm motion (question 8) and the location(s) of storm initiation (question 14). If you did not forecast the event well, what additional information may have helped you to make a better forecast?

_____

_____

_____

_____

_____

16. *(Advanced Students/Meteorology Majors)* Use the radar loop (displayed by your instructor) to comment on the evolution and motion of the storms that formed in both Comanche County and Major County (see Figure 9 for county names). In considering your answer, compare the motion of these two storms with the motion of other storms across Oklahoma.

_____

_____

_____

_____

_____

_____

Hurricane Tracks and Forecasts

LAB ACTIVITY OBJECTIVES:

- Given a series of hurricane locations expressed as pairs of latitude and longitude values, you will produce a plot of the track of a hurricane over time.
- You will study the symmetry or asymmetry of strong winds and high seas around the eye of a hurricane.
- Given a computer model forecast of the position and intensity of a hurricane or tropical storm, you will explore possible decisions to protect the lives of citizens.

MATERIALS NEEDED:

- Laboratory manual
- Pencil or pen
- Colored pencils or pens

GLOSSARY:

Asymmetric   *Rainband*
Eye   *Saffir-Simpson Hurricane Scale*
Landfall   *Storm Surge*
Landfalling Hurricane   *Symmetric*
Onshore   *Tropical Storm*

BACKGROUND:

Landfalling hurricanes not only disrupt the lives of individuals in coastal communities, they also can endanger people who live hundreds of miles inland. Even a small hurricane can have dangerous winds more than 100 miles from the **eye** position. When a hurricane makes **landfall**, these strong winds cause considerable damage and frequently kill people and animals that do not seek adequate protection. These winds and wind-carried debris also knock down electrical lines, resulting in even more dangerous conditions, especially during rain. Hurricanes often disrupt electrical service over swaths that can be 100 miles wide or larger, making restoration of power difficult and time consuming (e.g, weeks or months). Wind-blown debris is one of the worst hazards associated with the strong winds from hurricanes, tornadoes, severe thunderstorms, and even dust storms.

In addition to damage caused by hurricane winds, considerable damage results from the **storm surge**. In the open ocean, sea-surface winds flow towards the center of the hurricane, pushing water towards the center. As the water begins to pile up, a current is established where the water spirals down into the ocean to a depth of about 200 feet and then flows away from the center of the storm. During landfall, the ocean floor blocks the sub-surface currents that carry away excess water. As a result, the water swells higher and is pushed **onshore** by intense winds. This is the storm surge. The most intense storm surge is located to the right of the eye, where the wind blows onshore (i.e., from the ocean to the land; Figure 1).

Flooding is another danger of a hurricane. The heavy rains associated with a hurricane are responsible not only for major flooding in areas where the storm initially strikes, but also can affect areas hundreds of miles inland from the location of hurricane landfall.

Figure 1 – Schematic of Onshore and Offshore Flow as a Hurricane Approaches Land

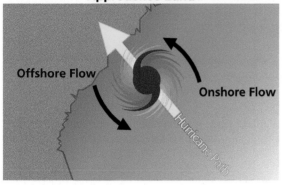

Offshore Flow

Onshore Flow

Hurricane Path

Weak tornadoes can occur with landfalling hurricanes. The tornado threat often is greatest in the right-front quadrant of the hurricane (i.e., along and to the right of the track of the eye). In addition, the tornado probability increases if the hurricane is configured such that the spiral rainbands advect dry air (from outside the hurricane) into its circulation. The juxtaposition of dry air and moist air across some of the **rainbands** can provide sufficient instability to spawn mini-supercell thunderstorms with associated updrafts and small tornadoes. In general, the formation of tornadoes within landfalling tropical systems is not well understood.

HURRICANE FORECAST AND WARNING

Traditionally, hurricanes are plotted using a latitude/longitude point that denotes the center of the hurricane circulation. Unfortunately, this method does not convey either the spatial extent or the intensity of the hurricane. To aid citizen's understanding of the severity of a particular hurricane, the **Saffir-Simpson Hurricane Scale**, based on maximum wind speeds, is used for both forecasts and warnings. Table 1 displays the Saffir-Simpson Hurricane Scale.

Although each hurricane has a unique wind field, storm surge, and rainfall pattern, the National Hurricane Center (NHC) produces forecast information to depict the hazards that accompany each hurricane. The NHC, located in Miami, FL, is an office within the National Weather Service (NWS) that has forecast and warning responsibility for the coastal areas of the United States during tropical events. The NHC issues watches and warnings for both hurricanes and **tropical storms**, including forecasts of landfall probabilities. The Storm Prediction Center (SPC) in Norman, OK, monitors the tornado threat of landfalling hurricanes and issues tornado watches when appropriate. Local NWS forecast offices provide local warnings, advisories, and forecasts of county-scale conditions, including flooding, high winds, and tornadoes. Local NWS offices also coordinate with local and state government agencies to help disseminate important information to citizens including evacuation routes and shelter locations.

Table 1 – Saffir–Simpson Hurricane Scale

Hurricane Category	Wind Speed (in mph)	Wind Speed (in km/hr)	Associated Damage
Category 1	74–95	119–153	Storm surge ~4–5 ft above normal; minor coastal flooding; wind damage to old or unsecured structures
Category 2	96–110	154–177	Storm surge ~6–8 ft above normal; flooding of low-lying coastal areas; some damage to windows and roofing; considerable damage to mobile homes
Category 3	111–130	178–209	Storm surge ~9–12 ft above normal; coastal and inland flooding of low-lying areas; trees downed, foliage stripped, mobile homes destroyed
Category 4	131–155	210–249	Storm surge ~13–18 ft above normal; significant flooding to 6 miles inland; trees and all signs downed, significant roof damage; mobile homes destroyed; extensive damage to windows and doors; significant damage to lowest floor of coastal structures
Category 5	156+	250+	Storm surge greater than 18 ft above normal; significant flooding to 10 miles inland; complete roof failure on many buildings; mobile homes destroyed; severe structural damage

Because hurricanes form and intensify over the ocean, one of the best tools to monitor their movement and development is the weather satellite. Visible satellite images are used to locate the center of the eye and to examine the structure and size of the hurricane. Infrared imagery, including water vapor images, are used to determine the height of clouds within the hurricane, to examine the flow of water vapor or dry air into the storm system, and to track the hurricane at night. Sequential images help to determine whether the hurricane is intensifying or weakening.

Figure 2 displays a visible satellite image of Hurricane Isabel at 1525 UTC on 15 September 2003. Isabel formed nine days earlier in the central Atlantic Ocean. Isabel was a Category–5 hurricane as she moved north of the Bahamas and Caribbean, and the storm made landfall on the U.S. East Coast as a Category–2 storm. In this laboratory exercise, you will examine the motion, intensity, and potential human impacts of Hurricane Isabel.

Figure 2 – Visible Satellite Image of Hurricane Isabel from 1525 UTC on 15 September 2003
(Courtesy of the Naval Research Laboratory)

LABORATORY EXERCISES:

1. Record the track of Hurricane Isabel by plotting the latitude/longitude pairs listed in Table 2 for 0300 UTC on 14 September 2003 to 1500 UTC on 17 September. Plot the locations on the visible satellite image (Figure 3). Note that the latitude and longitude gridlines on Figure 3 are 1° by 1°, so you will need to interpolate the positions to produce an accurate track.

Table 2 – Track of Hurricane Isabel for 36 Hours to 1500 UTC on 15 September 2003 and Forecast Track Positions to 48 Hours Based on Current Conditions

	Date and Time	Time	Latitude	Longitude	Intensity
36 Hours Ago	14 September	0300 UTC	23.0° N	63.7° W	140 kt
24 Hours Ago	14 September	1500 UTC	23.7° N	66.3° W	135 kt
12 Hours Ago	15 September	0300 UTC	24.5° N	68.3° W	135 kt
Current	15 September	1500 UTC	25.2° N	69.4° W	120 kt
12–Hour Forecast	16 September	0300 UTC	25.8° N	70.3° W	115 kt
24–Hour Forecast	16 September	1500 UTC	27.0° N	71.2° W	115 kt
36–Hour Forecast	17 September	0300 UTC	28.2° N	72.0° W	115 kt
48–Hour Forecast	17 September	1500 UTC	30.0° N	73.0° W	115 kt

Figure 3 – Visible Satellite Image of Hurricane Isabel from 1525 UTC on 15 September 2003 with Latitude and Longitude Gridlines
(Courtesy of the Naval Research Laboratory)

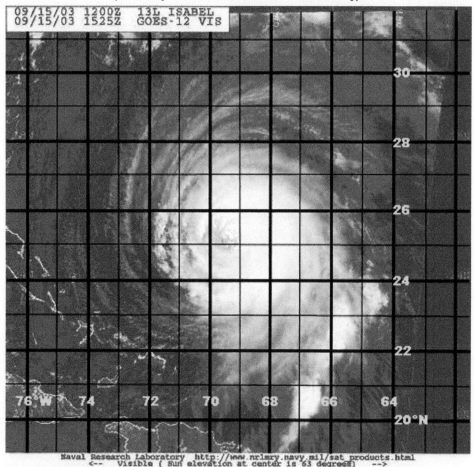

2. The National Hurricane Center expresses various hurricane parameters in terms of a radius from the hurricane center. For example, the NHC might state that 50–knot winds extend 60 miles northwest of the center. To plot the parameters on a standard map using a latitude/longitude grid, it is necessary to convert distances in miles (mi) to degrees latitude and degrees longitude. In the tropics, one degree of latitude or longitude equals ~70 statute miles. At higher latitudes, the spatial extent of one degree of longitude decreases. The spatial extent of one degree latitude is consistent everywhere. For the purposes of this lab, however, use 1 deg lat/lon = 70 miles.

50 mi = _____ deg lat/lon   180 mi = _____ deg lat/lon

100 mi = _____ deg lat/lon   225 mi = _____ deg lat/lon

140 mi = _____ deg lat/lon   575 mi = _____ deg lat/lon

150 mi = _____ deg lat/lon

3. On Figure 4, plot the center of Hurricane Isabel using the latitude and longitude of the "current" hurricane position listed in Table 2. Use your answers above and Table 3 to plot the radii of the 64-knot (with a red pencil), 50-knot (with a green pencil), and 34-knot (with a purple pencil) winds on Figure 4.

Table 3 – Wind and High Seas Radii Associated with Hurricane Isabel

Current Conditions – 15 September 2003 at 1500 UTC			
Hurricane Parameter	Radius to NE (in miles)	Radius to SE (in miles)	Radius to SW (in miles)	Radius to NW (in miles)
64 kt winds	100	100	100	100
50 kt winds	140	140	140	140
34 kt winds	180	180	180	180
12 ft seas	75	150	150	575

4. Examine the radii you drew on Figure 4. Are these winds **symmetric** or **asymmetric** about the center of Isabel?

_____

5. Do the radii of 34-knot winds (Figure 4) extend beyond the cloud cover noted in the image?

_____

6. Use Table 3 to plot the radii of the 12-foot seas (with a blue pencil) on Figure 4 from the "current" hurricane position listed in Table 3.

7. Explain why the radii of the 12-foot seas are <u>not</u> symmetric about the eye of the hurricane.

_____

_____

_____

**Figure 4 – Visible Satellite Image of Hurricane Isabel from
1525 UTC on 15 September 2003 with Latitude and Longitude Gridlines
(Courtesy of the Naval Research Laboratory)**

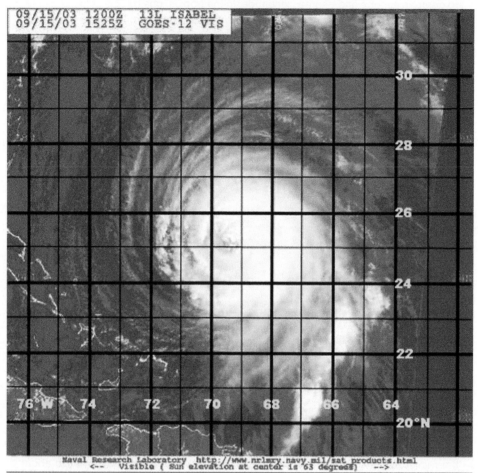

8. Using Figure 4, do the radii of the 12-foot seas extend beyond the cloud cover noted in the image?

_____

9. On Figure 5, plot the center of Hurricane Isabel using the latitude and longitude of the 48-hour forecast of hurricane position listed in Table 2. Use Table 4 and Figure 5 to plot the radii of the 48-hour forecast position of the 64-knot (with a red pencil), 50-knot (with a green pencil), and 34-knot (with a purple pencil) winds.

Table 4 – Forecasted Wind Radii Associated with Hurricane Isabel

Forecast – Valid 17 September 2003 at 1500 UTC			
Hurricane Parameter	Radius to NE (in miles)	Radius to SE (in miles)	Radius to SW (in miles)	Radius to NW (in miles)
64 kt winds	100	100	50	50
50 kt winds	150	150	100	100
34 kt winds	225	225	150	150

**Figure 5 – Visible Satellite Image of Hurricane Isabel from
1525 UTC on 15 September 2003 with Latitude and Longitude Gridlines
(Courtesy of the Naval Research Laboratory)**

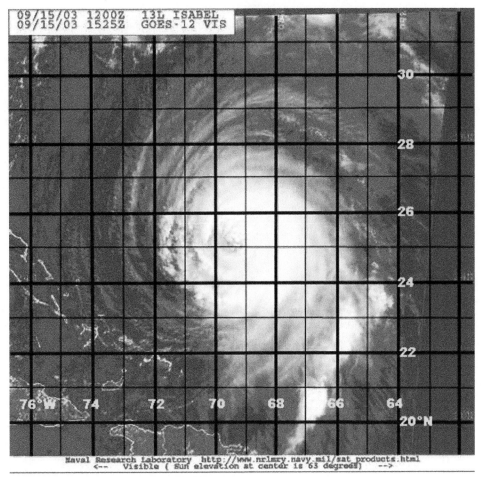

10. Examine the radii you drew on Figure 5. Are these winds symmetric or asymmetric about the center of Isabel?

_____

11. What factors might explain the change in symmetry between the wind radii you plotted in Figures 4 and 5?

_____

_____

_____

_____

_____

12. The Geophysical Fluid Dynamics Laboratory in Princeton, NJ, developed a hurricane forecast model that provides a forecast of position and intensity of a hurricane or tropical storm up to 84 hours (3.5 days) in advance. Figure 6 displays this model's 84-hour forecast valid for 8:00 PM EDT on 18 September 2003 (0000 UTC on 19 September 2003). The model placed the center of Tropical Storm Isabel (downgraded from hurricane status) over southern Virginia. Use Figure 6 to answer the following scenario:

It is Monday, September 15, 2003. You are the emergency manager for the City of Blacksburg, VA (located at the black dot on Figure 6). The Virginia Tech–Texas A&M football game in Blacksburg will be televised nationally on Thursday night, September 18. A sell-out crowd of 65,115 is expected in the stadium. Kickoff is set for 7:00 PM EDT. You are handed Figure 6. What meteorological and societal factors must you consider when deciding to possibly cancel, reschedule, or move the game? What decision would you make regarding the football game?

_____

_____

_____

_____

Figure 6 – 84–Hour Forecast Position and Intensity of Hurricane Isabel Valid at 0000 UTC on 19 September 2003 (8:00 PM EDT on 18 September 2003) (Courtesy of the National Hurricane Center)

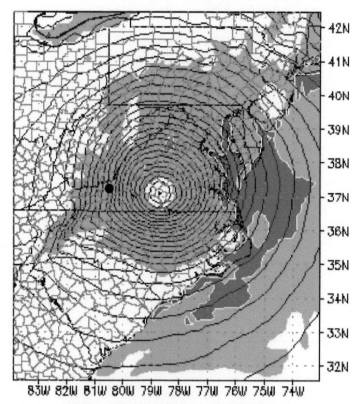

Min. Pressure = 965.9 mb
Max. Wind @ 35m = 63.4 kt

Weather Forecasting: Precipitation Type

LAB ACTIVITY OBJECTIVES:

- Using a Skew–T diagram, you will locate the precipitation generation region and deduce the phase of water (liquid drops or ice crystals) in this region.
- You will draw the wet-bulb temperature profile for the lower portion of a variety of atmospheric soundings.
- Using data from a variety of soundings, you will determine how precipitation changes as it falls through the cloud to the ground and diagnose the likely precipitation type at the ground.
- You will determine the precipitation type using the freezing-layer depth method and the 1000–500 mb thickness method and compare the pros and cons of these methods with your prior predictions of precipitation type.

MATERIALS NEEDED:

- Laboratory manual
- Pencil or pen
- Colored pencils

GLOSSARY:

Dry Adiabat	Moist Adiabat
Freezing Rain	Sleet
Ice Pellets	Supercooled Liquid Water
Lifting Condensation Level (LCL)	Surface Layer
Mesoscale	Thickness
Mixing Ratio	Wet–Bulb Temperature

BACKGROUND:

Freezing precipitation can result in disastrous consequences, including multi-vehicle accidents and extensive power outages. Unfortunately, freezing precipitation is one of the most difficult weather events for meteorologists to forecast. Like thunderstorms, the **mesoscale** nature of ice and snow storms often make them difficult to diagnose using the conventional radiosonde network. Forecasters must be aware of the processes that combine to produce freezing precipitation.

Perhaps the most critical and difficult component of a winter weather forecast is determining the vertical structure of temperature and moisture. These vertical profiles can be complicated, yet they generate the various forms of freezing (**freezing rain**), frozen (**sleet**, snow), and non-frozen (rain) precipitation. The most challenging forecasts often involve a combination of frozen ground, near-surface temperatures that are below freezing, and a shallow layer above the surface (e.g., from 800 mb to 900 mb) that is above freezing. A forecast error of only a few degrees Celsius can mean the difference between a forecast of rain, freezing rain, or snow.

VERTICAL PROFILE OF WET–BULB TEMPERATURE

The evaporation of falling precipitation can play an important role in cooling the **surface layer** sufficiently to change a rain event to a freezing rain event. Hence, to forecast how precipitation events will progress, it is important to determine the vertical profile of **wet-bulb temperature** from the latest radiosonde observations.

The wet-bulb temperature is the lowest temperature to which air can be cooled by evaporating moisture into the air parcel at constant pressure. The wet-bulb temperature is never greater than the temperature and never less than the dew point. If the temperature equals the dew point, then the wet-bulb temperature also will equal the temperature and dew point. The wet-bulb temperature can be calculated graphically using the following process on the Skew-T diagram:

1. To determine the wet-bulb temperature at a given level, first find the dewpoint temperature at the same level. Draw a line upward, parallel to the **mixing ratio** lines.

2. Find the temperature at the original, given level. Draw a line upward, parallel to the **dry adiabats**.

3. Find the point where these two lines intersect (i.e., the **LCL**). Draw a line downward, parallel to the **moist adiabats**, until it intersects the original pressure level. The temperature at the point of intersection is the wet-bulb temperature.

WINTER PRECIPITATION PROCESSES

Winter season precipitation typically generates from ice crystals within a cloud where wet-bulb temperatures are between –10°C and –15°C. As these ice crystals fall, they will melt if they encounter a sufficiently deep layer of air with temperatures warmer than 0°C. If the wet-bulb temperature of the warm layer is greater than 3°C, the ice crystals will melt and likely will not refreeze before hitting the surface. If the wet-bulb temperature of the warm layer does not exceed 1°C, then very little melting will occur. Varying degrees of melting occur when the wet-bulb temperature in the warm layer is between 1°C and 3°C.

Freezing Rain

If the ice crystals melt in the warm layer, the precipitation falls as liquid drops toward the ground. If a layer of subfreezing temperatures (T < 0°C) is encountered, the liquid drops may freeze on impact with a frozen surface. This type of precipitation is called freezing rain. Freezing rain also can occur if the surface is above freezing and the air next to the surface is dry enough such that the wet-bulb temperature is less than 0°C. In this situation, evaporational cooling will make the air next to the surface subfreezing, and freezing rain again will result.

Sleet

If the warm layer did not permit complete melting of the descending ice crystals, or if the cold surface layer was deep enough (> 1 km deep) to permit the liquid drops to refreeze before impact with the surface, then sleet or **ice pellets** result.

Snow and Rain

Snow results if no melting of ice crystals occurs (i.e., the air temperature is below freezing throughout the entire layer from the ice crystal generation region to the surface). The precipitation is rain, however, if the ice crystals melt and do not refreeze (i.e., both surface and wet-bulb temperatures are warmer than 0°C).

Winter Precipitation Mixture

As discussed above, the temperature structure of the lower atmosphere is absolutely critical in generating a particular type of precipitation. The above scenarios also assumed that the precipitation particles began their descent as ice crystals. Another complication to forecasters is that *liquid* water drops can exist in clouds with temperatures below 0°C. These drops are called **supercooled liquid water**. This scenario results if precipitation is generated in a cloud regime where the wet-bulb temperature of the cloud layer is warmer than –10°C. In this case, the precipitation particles likely will originate as supercooled liquid and they could reach the surface as either rain, sleet, or freezing rain (but not snow).

Tables 1 and 2 summarize the threshold values of wet bulb temperature for determining precipitation type. The values in Table 1 indicate whether the precipitation originates as liquid or as ice crystals. Precipitation initiated as supercooled liquid water results in rain (if the surface is above freezing), freezing rain (if the precipitation falls through a warm layer but the surface is below 0°C), or sleet (if the precipitation descends through a layer sufficiently deep and cold for the supercooled drops to freeze). Table 2 presents the surface precipitation types for clouds composed of ice crystals. The precipitation type at the surface depends upon both the surface temperature and the temperature profile between the surface and the precipitation generation region.

Table 1 – Wet–Bulb Temperature Thresholds in the Precipitation Generation Region

Wet–Bulb Temperature	Cloud Precipitation Particle Type
-10°C to +1°C	Supercooled Liquid Water
< -10°C	Snow/Ice Crystals

Table 2 – Wet–Bulb Temperature Thresholds in the Precipitation Fall Region When the Precipitation Is Generated as Ice Crystals

Wet–Bulb Temperature	Surface Precipitation Type
>3°C	Rain, Freezing Rain if surface T is < 0°C
+1°C to +3°C	Mixed (Sleet, Freezing Rain or Snow)*
< 0°C	Snow

*The type of mixed precipitation depends upon the depth and the wet-bulb temperature of the precipitation fall region.

Figure 1 – Schematic Diagram Demonstrating How to Compute the Wet–Bulb Temperature

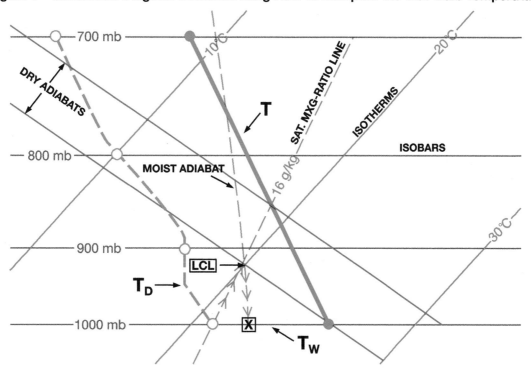

OTHER METHODS TO DETERMINE PRECIPITATION TYPE

Because of the complexities involved with determining precipitation type directly from soundings, forecasters sometimes use two simplified approaches: the freezing-layer depth method and the 1000–500 mb **thickness** method. Similar to using stability indices from soundings for severe weather predictions, the simplified methods often overlook key details and may result in significant forecast errors.

Freezing-Layer Depth Method

Because the freezing layer must be deep enough to keep freezing precipitation frozen, forecasters often estimate the depth of the freezing layer from 850-mb or 700-mb temperature maps. If the 850-mb temperature is below freezing, forecasters often reason that the air below 850 mb also is subfreezing and the precipitation will fall as snow. In regions of higher terrain, 850 mb may occur at or near the surface, so forecasters may use the 700 mb level instead.

1000–500 mb Thickness Method

Forecasters often look at the 1000–500 mb thickness to classify regions according to precipitation type. In practice, regions with thickness values less than 540 decameters (dam) will receive snow; regions with thickness values near 540 dam will experience mixed precipitation; and regions with thickness values greater than 540 dam will receive rain (Table 3). The threshold value of 540 dam is based on statistical studies of freezing precipitation events that occurred east of the Rocky Mountains. The 1000–500 mb thickness (Z) is calculated simply as $Z = z$(at 500 mb) $- z$(at 1000 mb), where z is the height computed from the sounding.

Table 3 – 1000–500 mb Thickness and Associated Surface Precipitation Type

1000–500 mb Thickness	Surface Precipitation Type
>540 dam	Rain
~540 dam	Mixed (Sleet, Freezing Rain or Snow)
< 540 dam	Snow

LABORATORY EXERCISES:

You will use three different soundings to determine the type of precipitation (e.g., ice crystal, liquid) in the precipitation generation region. Figure 2 and Table 4 display sounding data from Little Rock, AR (LZK) for 23 December 1998 at 1200 UTC. Figure 3 and Table 5 show the data from Fort Worth, TX (FWD) for the same date and time. Figure 4 and Table 6 display radiosonde data from Detroit, MI (DTX) for 3 January 1999 at 0000 UTC. Before you begin the questions, study the soundings as you would if you were preparing a weather forecast.

1. To identify on a Skew–T diagram where precipitation is generated in the atmosphere above the radiosonde site, locate any deep layer where the air is saturated. [Note: Air can be saturated when the dew point measured at a given level is within 2–3°C of the temperature at that level.] Find the top of the highest saturated layer (i.e., saturated layer farthest from the ground) and list the pressure at this height for LZK, FWD, and DTX.

    LZK _____mb   FWD _____mb   DTX _____mb

2. What is the temperature at the pressure level that you identified in question #1?

    LZK _____°C   FWD _____°C   DTX _____°C

3. What is the wet-bulb temperature (T_w) at the pressure level identified in question #1?

    LZK _____°C   FWD _____°C   DTX _____°C

4. Identify the phase of water located at the precipitation generation regions that were identified in question 1. Specifically, is the water in liquid (i.e., $T_w > -10°C$) or ice crysal ($T_w < -10°C$) form?

                    LZK = liquid or ice crystals (circle one)

                    FWD = liquid or ice crystals (circle one)

                    DTX = liquid or ice crystals (circle one)

5. For each of the three locations (i.e., LZK, FWD, and DTX), determine what happens to the precipitation particles as they fall toward the ground. Discuss the precipitation particles and their journey to the surface. Include in your discussion whether the particles melt, freeze, experience evaporational cooling, etc. To help with your answer, calculate the wet-bulb temperature profile for the area of interest (from the precipitation generation area down to the surface) and draw it on the soundings in Figures 2, 3, and 4.

    a. Little Rock, AR (LZK):

    _____

    _____

    _____

    _____

    _____

    _____

Figure 2 – Skew T–Log P Plot of Sounding Data from Little Rock, AR (LZK) for 1200 UTC on 23 December 1998

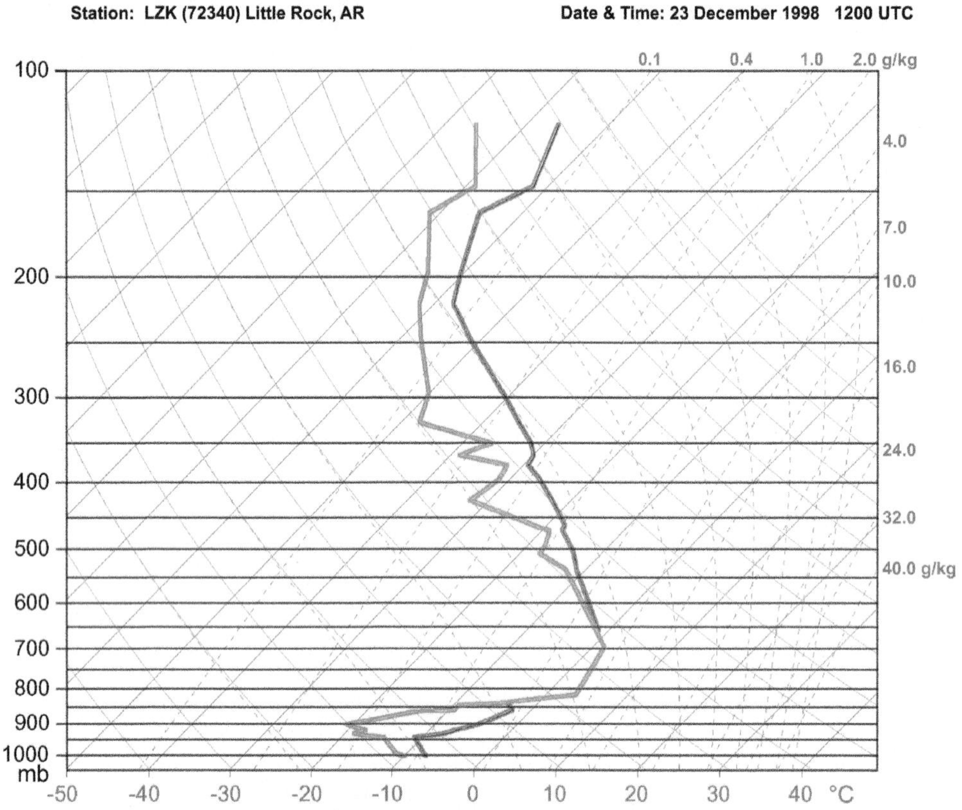

Station: LZK (72340) Little Rock, AR   Date & Time: 23 December 1998   1200 UTC

Table 4 – Sounding Data from Little Rock, AR (72340 LZK) for 23 December 1998 at 1200 UTC

P (mb)	Z (m)	T (°C)	T$_D$ (°C)	RH (%)	w (g/kg)	WDIR (deg)	WSPD (kts)
1009.0	78	−7.5	−10.1	82	1.77	30	10
1000.0	**243**	**−8.1**	**−11.4**	**77**	**1.61**	**35**	**12**
992.0	305	−8.6	−11.9	77	1.56	40	12
953.6	610	−10.8	−14.4	74	1.32	55	14
948.0	656	−11.1	−14.8	74	1.29	55	13
936.0	754	−7.9	−18.9	41	0.92	55	11
925.0	**846**	**−6.9**	**−17.9**	**41**	**1.02**	**55**	**10**
917.0	914	−6.0	−19.2	34	0.92	60	12
907.0	1000	−4.9	−20.9	27	0.80	46	11
882.1	1219	−3.5	−16.3	36	1.22	10	8
868.0	1346	−2.7	−13.7	43	1.54	342	8
864.0	1383	−2.5	−9.5	59	2.16	334	8
850.0	**1512**	**−3.5**	**−9.5**	**63**	**2.20**	**305**	**8**
845.0	1559	−3.5	−4.6	92	3.23	297	9
837.0	1634	−2.7	−2.7	100	3.76	284	11
822.0	1779	3.4	3.4	100	5.98	259	14
817.0	1829	3.3	3.3	100	5.98	250	16
786.8	2134	2.8	2.8	100	5.99	235	29
757.8	2438	2.3	2.3	100	5.99	240	35
729.8	2743	1.8	1.8	100	5.99	250	43
702.8	3048	1.2	1.2	100	6.00	255	47
700.0	**3080**	**1.2**	**1.2**	**100**	**6.00**	**255**	**49**
650.4	3658	−2.3	−2.7	97	4.84	250	54
602.0	4267	−6.1	−6.8	94	3.83	245	60
557.1	4877	−9.8	−10.9	91	2.99	245	68
540.0	5123	−11.3	−12.6	90	2.70	244	68
512.0	5530	−13.5	−17.6	71	1.88	241	68
500.0	**5710**	**−14.7**	**−17.8**	**77**	**1.90**	**240**	**68**

Figure 3 – Skew T–Log P Plot of Sounding Data from Fort Worth, TX (FWD) for 1200 UTC on 23 December 1998

Station: FWD (72249) Fort Worth, TX   Date & Time: 23 December 1998 1200 UTC

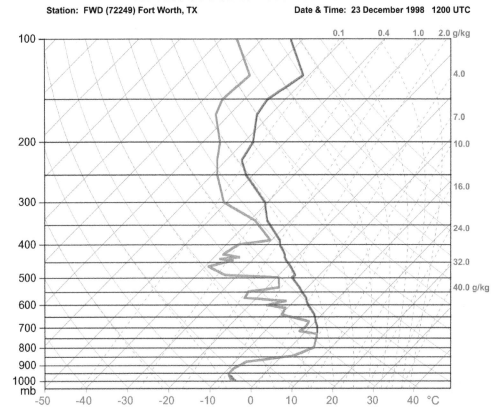

Table 5 – Sounding Data from Ft. Worth, TX (72249 FWD) for 23 December 1998 at 1200 UTC

P (mb)	Z (m)	T (°C)	T$_D$ (°C)	RH (%)	w (g/kg)	WDIR (deg)	WSPD (kts)
1005.0	171	−4.5	−4.8	98	2.67	80	4
1000.0	**237**	**−4.9**	**−5.7**	**94**	**2.51**	**89**	**4**
963.0	532	−7.5	−7.6	99	2.25	159	8
925.0	**846**	**−7.7**	**−7.7**	**100**	**2.33**	**234**	**11**
917.0	914	−7.4	−7.4	100	2.40	250	12
885.0	1191	−6.3	−6.3	100	2.71	255	15
881.9	1219	−5.4	−5.4	100	2.91	255	16
850.0	**1513**	**3.8**	**3.8**	**100**	**5.95**	**250**	**25**
848.9	1524	3.9	3.9	100	5.98	240	33
817.8	1829	5.8	5.8	100	7.14	230	47
803.0	1978	6.8	6.8	100	7.77	228	48
787.9	2134	6.3	6.3	100	7.65	225	49
759.2	2438	5.3	5.3	100	7.41	225	47
734.0	2715	4.4	4.4	100	7.19	230	38
731.4	2743	4.3	3.5	95	6.77	230	37
720.0	2871	3.8	−0.6	73	5.11	230	34
704.5	3048	3.0	−0.2	79	5.38	230	31
700.0	**3100**	**2.8**	**−0.1**	**81**	**5.46**	**230**	**35**
674.0	3406	1.0	−0.6	89	5.46	233	38
653.2	3658	−0.3	−6.3	64	3.67	235	41
644.0	3771	−0.9	−8.9	55	3.04	236	41
616.0	4125	−3.5	−9.5	63	3.04	239	42
605.0	4267	−4.6	−14.3	47	2.11	240	43
604.0	4280	−4.7	−14.7	46	2.03	240	43
587.0	4504	−6.1	−11.1	68	2.81	240	45
575.0	4666	−6.9	−21.9	29	1.16	240	47
559.6	4877	−8.6	−22.3	32	1.15	240	49
551.0	4997	−9.5	−22.5	34	1.15	241	50
535.0	5224	−11.1	−16.0	67	2.06	242	54
500.0	**5740**	**−15.3**	**−18.5**	**76**	**1.79**	**245**	**62**

Figure 4 – Skew T–Log P Plot of Sounding Data from Detroit, MI (DTX) for 0000 UTC on 3 January 1999

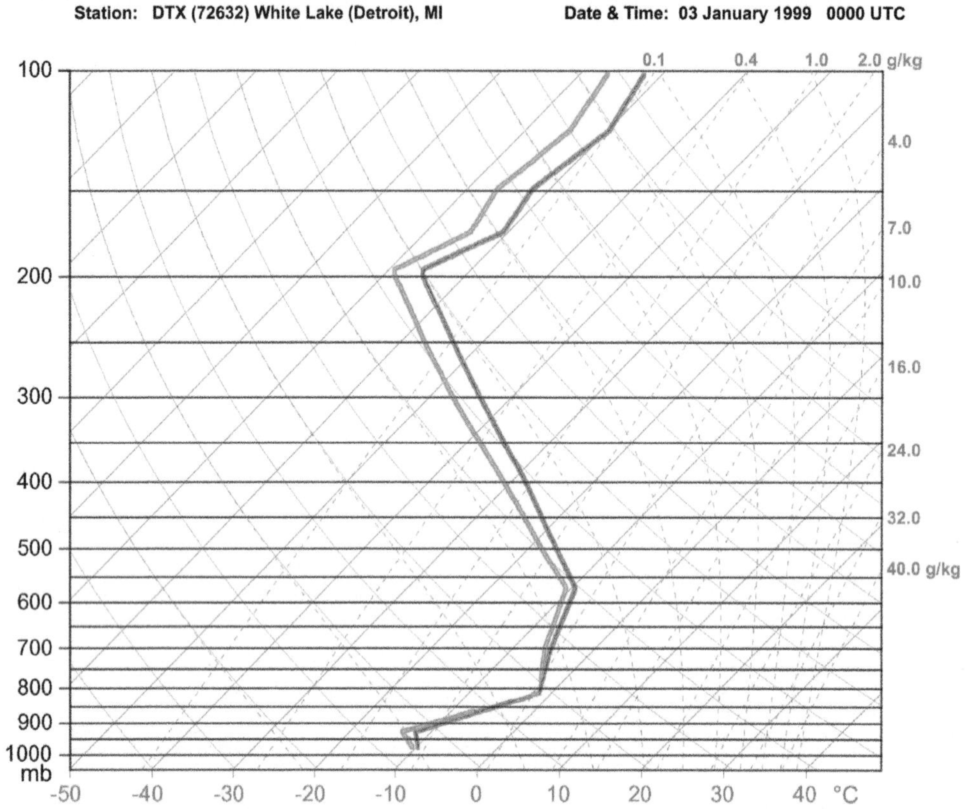

Station: DTX (72632) White Lake (Detroit), MI   Date & Time: 03 January 1999 0000 UTC

Table 6 – Sounding Data from Detroit, MI (72632 DTX) for 3 January 1999 at 0000 UTC

P (mb)	Z (m)	T (°C)	T_D (°C)	RH (%)	w (g/kg)	WDIR (deg)	WSPD (kts)
1000.0	**109**						
972.0	329	−10.1	−10.8	95	1.73	90	25
936.9	610	−11.6	−12.9	90	1.51	100	49
925.0	**708**	**−12.1**	**−13.7**	**88**	**1.44**	**105**	**49**
900.7	914	−10.1	−11.4	90	1.79	125	45
866.0	1219	−7.1	−7.9	94	2.44	150	45
850.0	**1363**	**−5.7**	**−6.3**	**96**	**2.82**	**160**	**43**
812.0	1725	−1.9	−1.9	100	4.12	187	44
801.3	1829	−2.2	−2.3	100	4.05	195	45
771.0	2134	−3.2	−3.5	98	3.86	210	43
741.9	2438	−4.2	−4.6	97	3.67	215	41
713.9	2743	−5.2	−5.8	95	3.49	210	39
700.0	**2898**	**−5.7**	**−6.4**	**95**	**3.40**	**205**	**41**
635.0	3658	−7.7	−8.6	93	3.16	205	72
587.3	4267	−9.3	−10.4	92	2.97	200	89
571.0	4487	−9.9	−11.0	92	2.91	198	89
542.5	4877	−12.6	−14.0	89	2.41	195	87
500.0	**5500**	**−16.9**	**−18.7**	**86**	**1.76**	**195**	**85**

5. (Continued)
    b. Fort Worth, TX (FWD):

    _____

    _____

    _____

    _____

    _____

    _____

    c. Detroit MI (DTX):

    _____

    _____

    _____

    _____

    _____

    _____

6. For each of the three soundings, identify the 850-mb temperatures (in °C) using Tables 4, 5, and 6. Based on this information alone, what precipitation type would you expect at the surface? Do these expectations agree with your conclusions above? Enter your answers in Table 7.

Table 7 – 850-mb Temperature and Precipitation Type for LZK, FWD, and DTX

Station	850-mb Temp. (in °C)	Precipitation Type	Agreement? (Y/N)
Little Rock, AR (LZK)			
Ft. Worth, TX (FWD)			
Detroit, MI (DTX)			

7. Explain any pros and cons of using the 850-mb temperature method from question 6 to estimate the precipitation type.

    _____

    _____

    _____

8. For each of the three soundings, calculate the 1000–500 mb thickness (in meters) using the information in Tables 4, 5, and 6. Based on this information alone, what precipitation type would you expect at the surface? Do these expectations agree with your conclusions above? Enter your answers in Table 8.

Table 8 – 1000–500 mb Thickness and Precipitation Type for LZK, FWD, and DTX

Station	1000–500 mb Thickness (in m)	Precipitation Type	Agreement? (Y/N)
Little Rock, AR (LZK)			
Ft. Worth, TX (FWD)			
Detroit, MI (DTX)			

9. Explain any pros and cons of using the thickness method from question 8 to estimate the precipitation type.

_____

_____

_____

_____

_____

_____

_____

10. *(Advanced Students/Meteorology Majors)* Explain the difference between the dewpoint temperature and the wet-bulb temperature.

_____

_____

_____

_____

_____

_____

_____

_____

Weather Forecasting: Model Output Statistics

LAB ACTIVITY OBJECTIVES:

- You will examine Model Output Statistics (MOS) from two different models to analyze forecast bias for maximum temperature, minimum temperature, and probability of precipitation.
- Using MOS output and actual observations, you will make inter-comparisons between forecasts from different models to assess forecast accuracy.
- You will compare the accuracy of 60-hour forecasts with the corresponding 24-hour forecasts.

MATERIALS NEEDED:

- Laboratory manual
- Pencil or pen

GLOSSARY:

Forecast Accuracy
Forecast Bias
Global Forecast System (GFS)
Model Output Statistics (MOS)
Nested Grid Model (NGM)
Numerical Weather Prediction (NWP)
Probability of Precipitation (POP)

BACKGROUND:

The earth-atmosphere system encompasses a planet with vast oceans, diverse vegetation, and uneven terrain, enveloped by a 40-kilometer-deep mixture of gases. Within the atmosphere, concentrations of gases vary both temporally and spatially, and water exists in vapor, liquid, and solid forms. Beneath the atmosphere, the earth's surface heats and cools irregularly, and plants, lakes, and oceans add moisture to the air. Beyond the atmosphere, the sun radiates energy from 150 million kilometers away, unevenly heating the spherical earth that rotates on its 23.5-degree axis. Each component of this earth-atmosphere system reacts differently to the inputs of heat and moisture, resulting in a myriad of weather and climate patterns across the globe. Yet even with these variables and more, meteorologists are expected to provide accurate weather forecasts 24 hours per day, 7 days per week, for all locations where people work or play.

For centuries, weather forecasts were based upon lore, personal observations, recurring atmospheric phenomena, atmospheric optics, human health, clouds, or animal behavior. During those days, weather forecasting was more of an art than a science. During the Renaissance Age, however, great thinkers realized that further understanding of the earth-atmosphere system would require detailed measurements of atmospheric properties such as moisture, temperature, and pressure. Notable figures in colonial America, including Benjamin Franklin and Thomas Jefferson, meticulously recorded weather observations and began to produce simple maps and forecasts. From these observations and others, theories describing atmospheric motion were developed and expressed using mathematical equations.

Numerical Weather Prediction (NWP)

In 1904, the idea of predicting the weather by solving mathematical equations was formulated by Vilhelm Bjerknes and advanced by Lewis Fry Richardson, a British mathematician. During the late 1940s, mathematician John von Neumann directed the construction of a computer that was used to predict weather. He and Jule Charney determined that many of the problems in weather forecasting could be overcome by using computers and a simplified set of equations focused on synoptic-scale systems. Based on physical and dynamical processes, these equations use an initial set of surface, sounding, and other atmospheric observations, and advance them forward in time to produce a forecast. In April 1950, Charney's group made a series of successful 24-hour forecasts across North America. This **numerical weather prediction (NWP)** has become an integral part of modern forecasting.

Model Output Statistics (MOS)

In 1972, Harry Glahn and Dale Lowry outlined an objective forecast technique that is used to enhance **forecast accuracy**. This technique, called **Model Output Statistics (MOS)**, uses a statistical relationship that correlates the output from a specific NWP model with traditional variables such as temperature, **probability of precipitation (POP)**, wind speed, and wind direction. The correlations are site-specific and, hence, can enhance a standard computer prediction with the climatology of the site. In addition, the MOS forecasts remove many of the biases of a particular NWP model forecast by comparing past biases to real observations over a long period of time. MOS forecasts are used operationally in both government and private weather forecast offices across the U.S.

LABORATORY EXERCISES:

In this laboratory exercise, a limited set of Model Output Statistics (MOS) data from two popular numerical weather prediction models will be compared over a five-day period (Tables 1-6). The MOS data included in this lab are from two different operational models run by the U.S. National Weather Service. The models are called the **Nested Grid Model (NGM)** and the **Global Forecast System (GFS)**. These forecasts include maximum temperature, minimum temperature, and the probability of precipitation for Williamsport, PA. This lab can serve as a model for any city, any time period, and any MOS forecast variable.

To read Tables 1, 2, 4, and 5, first find the date and time that the model forecast was initialized (i.e., the date and time of the most recent observations that were input into the model). For example, in Table 1, find the row on the left side where the NGM MOS initialized its forecast with data from 1200 UTC on 6 December (12/6). Read across the row to find the 24-, 36-, 48-, and 60-hour forecasts. In this case, the 24-hour forecast (shaded in pink) for the 1200 UTC run on 6 December projects the minimum temperature for 7 December to be 36°F. The 36-hour (shaded in light green) and 48-hour (shaded in light blue) forecasts predict the maximum and minimum temperatures for 8 December to be 47°F and 42°F, respectively. The 60-hour forecast (shaded in gold) of the same model run predicts the minimum temperature for 9 December to be 48°F.

Part I – Forecast Accuracy by Variable

1. Using only the *maximum temperature* forecasts (values in dark red) from the NGM MOS in Table 1 and from the GFS MOS in Table 2, discuss the differences in forecasts for the same time periods. (Note: Each column represents the same time period.) In your discussion, note if one model appears to be consistently warmer or cooler than the other (i.e., a **forecast bias**).

   _____

   _____

   _____

   _____

2. Using only the minimum *temperature* forecasts (values in dark blue) from the NGM MOS in Table 1 and from the GFS MOS in Table 2, discuss the differences in forecasts for the same time periods. In your discussion, note if one model appears to be consistently warmer or cooler than the other.

   _____

   _____

   _____

3. Forecasts are verified using actual measurements. Using the observations in Table 3 for verification, which model was more accurate in its forecast of maximum temperature during this limited time period? You can compute the difference between the forecasted temperature and the actual temperature to represent the accuracy of the forecast. (Note: In most cases, the *square* of the difference is used to represent forecast accuracy.)

   _____

   _____

   _____

   _____

Table 1 – 24-, 36-, 48-, and 60-Hour Temperature (Max/Min) Forecasts for Williamsport, PA from the Model Output Statistics of the Nested Grid Model (NGM MOS)

Forecast Initialized		Maximum and Minimum Temperatures (in °F) As Forecast for 12-Hour Period Beginning at Date & Time											
Month/ Date	Time (UTC)	12/6 0000	12/6 1200	12/7 0000	12/7 1200	12/8 0000	12/8 1200	12/9 0000	12/9 1200	12/10 0000	12/10 1200	12/11 0000	12/11 1200
12/5	0000	48	28	35	31								
12/5	1200		25	36	36	42							
12/6	0000			37	36	48	39						
12/6	1200				36	47	42	48					
12/7	0000					48	41	50	32				
12/7	1200						43	48	32	49			
12/8	0000							50	33	46	37		
12/8	1200								36	46	42	44	
12/9	0000									46	39	44	41

Table 2 – 24-, 36-, 48-, and 60-Hour Temperature (Max/Min) Forecasts for Williamsport, PA from the Model Output Statistics of the Global Forecast System (GFS MOS)

Forecast Initialized		Maximum and Minimum Temperatures (in °F) As Forecast for 12-Hour Period Beginning at Date & Time											
Month/ Date	Time (UTC)	12/6 0000	12/6 1200	12/7 0000	12/7 1200	12/8 0000	12/8 1200	12/9 0000	12/9 1200	12/10 0000	12/10 1200	12/11 0000	12/11 1200
12/5	0000	49	29	39	39								
12/5	1200		33	38	37	50							
12/6	0000			37	36	51	43						
12/6	1200				39	50	44	53					
12/7	0000					51	42	52	31				
12/7	1200						42	54	29	44			
12/8	0000							54	31	44	38		
12/8	1200								33	44	39	45	
12/9	0000									46	39	46	41

Table 3 – Observed Maximum and Minimum Temperatures for Williamsport, PA for Model Verification

Observed Maximum and Minimum Temperatures (in °F)											
12/6 0000	12/6 1200	12/7 0000	12/7 1200	12/8 0000	12/8 1200	12/9 0000	12/9 1200	12/10 0000	12/10 1200	12/11 0000	12/11 1200
53	28	40	39	47	36	54	30	44	45	37	46

4. Using the observations in Table 3 for verification, which model was more accurate in its forecast of minimum temperature during this limited time period?

_____

_____

_____

_____

5. Using only the probability of precipitation (POP) forecasts from the NGM in Table 4 and from the GFS in Table 5, discuss the differences in forecasts for the same time periods. In your discussion, note if one model appears to consistently produce greater or fewer chances of precipitation than the other.

_____

_____

_____

_____

Table 4 – 24-, 36-, 48-, and 60-Hour Probability of Precipitation (POP) Forecasts for Williamsport, PA from the Model Output Statistics of the Nested Grid Model (NGM MOS)

Forecast Initialized		Probability of Precipitation (in percent) As Forecast for 12-Hour Period Beginning at Date & Time											
Month/ Date	Time (UTC)	12/6 0000	12/6 1200	12/7 0000	12/7 1200	12/8 0000	12/8 1200	12/9 0000	12/9 1200	12/10 0000	12/10 1200	12/11 0000	12/11 1200
12/5	0000	0	18	56	22								
12/5	1200		2	67	53	95							
12/6	0000			71	61	91	17						
12/6	1200				77	96	43	0					
12/7	0000					91	52	0	3				
12/7	1200						71	6	0	9			
12/8	0000							13	0	7	43		
12/8	1200								2	45	93	79	
12/9	0000									50	96	100	63

Table 5 – 24-, 36-, 48-, and 60-Hour Probability of Precipitation (POP) Forecasts for Williamsport, PA from the Model Output Statistics of the Global Forecast System (GFS MOS)

Forecast Initialized		Probability of Precipitation (in percent) As Forecast for 12-Hour Period Beginning at Date & Time											
Month/ Date	Time (UTC)	12/6 0000	12/6 1200	12/7 0000	12/7 1200	12/8 0000	12/8 1200	12/9 0000	12/9 1200	12/10 0000	12/10 1200	12/11 0000	12/11 1200
12/5	0000	1	11	73	52								
12/5	1200		2	81	59	100							
12/6	0000			79	83	100	37						
12/6	1200				57	100	36	4					
12/7	0000					96	47	4	5				
12/7	1200						49	4	4	17			
12/8	0000							1	1	13	57		
12/8	1200								2	14	65	100	
12/9	0000									31	83	92	87

Table 6 – Observed 24-Hour Precipitation for Williamsport, PA for Model Verification

Measured Precipitation (in inches)						
12/6 0000 – 12/6 1200	12/6 1200 – 12/7 1200	12/7 1200 – 12/8 1200	12/8 1200 – 12/9 1200	12/9 1200 – 12/10 1200	12/10 1200 – 12/11 1200	12/11 1200 – 12/12 1200
N/A	0.08	0.36	0.00	0.48	0.85	0.01

6. Using the observations in Table 6 for verification, which model was more accurate in its forecast of precipitation during this limited time period?

_____

_____

_____

7. Discuss the difficulty of verifying a precipitation forecast when the predictor is the POP (probability of precipitation) and the observations used for validation is 24-hour total precipitation.

_____

_____

_____

_____

_____

_____

_____

Part II – Forecast Accuracy by Length of Forecast Period

In this section, the accuracy of 60-hour forecasts (shaded in gold in the tables) will be compared to that of 24-hour forecasts (shaded in pink in the tables). To make a comparison, find a 60-hour forecast (the last number in any given row). Find the corresponding 24-hour forecast by following the column downward to the bottom value. For example, in Table 1, 42°F is the 60-hour forecast of maximum temperature that is valid for the 12-hour period beginning at 0000 UTC on 8 December. This 60-hour forecast was initialized with observations from 5 December at 1200 UTC. The corresponding 24-hour forecast of maximum temperature that is valid for this same time is 48°F, from the forecast initialized with observations from 7 December at 0000 UTC.

8. Compare the accuracy of the 60-hour forecasts of *maximum temperature* of the NGM MOS (Table 1) and the GFS MOS (Table 2) with the corresponding 24-hour forecasts valid for the same time period. In particular, examine the forecasts valid for 0000 UTC on December 8, 9, and 10. Which of these two forecasts — 24-hour or 60-hour — is more accurate and by how much?

_____

_____

_____

_____

_____

9. Compare the accuracy of the 60-hour forecasts of *minimum temperature* of the NGM MOS (Table 1) and the GFS MOS (Table 2) with the corresponding 24-hour forecasts valid for the same time period. In particular, examine the forecasts valid for 1200 UTC on December 7, 8, and 9. Which of these two forecasts — 24-hour or 60-hour — is more accurate and by how much?

_____

_____

_____

_____

_____

10. Compare the accuracy of the 60-hour forecasts of *precipitation* of the NGM MOS (Table 4) and the GFS MOS (Table 5) with the corresponding 24-hour forecasts valid for the same time period. In particular, examine the forecasts valid for 0000 UTC on December 8, 9, and 10. Which of these two forecasts — 24-hour or 60-hour — is more accurate and by how much?

_____

_____

_____

_____

_____

Climate Statistics

LAB ACTIVITY OBJECTIVES:

- Given a multi-day or multi-year dataset, you will calculate and interpret various climate statistics (e.g., mean, median) for temperature and rainfall variables.
- Given a 100-year dataset, you will discuss the differences in the climate statistics of the full period-of-record and a 30-year subset of the record.
- Using a 30-year dataset, you will calculate and interpret a trendline using linear regression.
- You will discuss the factors that can account for differences in the shapes of the frequency distributions of temperature and precipitation at a given observation station.
- Given a dataset and corresponding graph of the probability of exceedence, you will examine the return period for precipitation at the station.

MATERIALS NEEDED:

- Laboratory manual
- Pencil or pen
- Calculator

GLOSSARY:

Climate	*Median*	*Sequence*
Climate Change	*Mode*	*Series*
Climate Variability	*Normal*	*Standard Deviation*
Covariance	*Probability Distribution*	*Statistic*
Cumulative Distribution Function	*Probability Distribution Function*	*Trend*
Frequency Distribution	*Probability of Exceedence*	*Trendline*
Gaussian	*Probability of Occurrence*	*Variance*
Linear Regression	*Range*	*Weather*
Mean	*Return Period*	

BACKGROUND:

Climate is a characterization of weather patterns during an extended period. The description of a region's climate is the result of observations of specific variables, such as temperature and precipitation, throughout a long period of time (e.g., at least 30 years). Although long-term averages often are used to describe the climate, it is important to understand that climate consists of a range of possible weather conditions. For this reason, climate typically is described using **statistics**, representing varying likelihoods, or probabilities, of weather conditions.

Climate summarizes the individual daily weather patterns over a given time. As a result, it is incorrect to interpret any single weather event as representative of the climate. For example, a climatological drought occurs over several months or longer. One particular rainfall event, even of several inches of rain, does not necessarily signal the end of the drought. Many individual rainfall events may be required to end the drought.

CLIMATE STATISTICS

Climate statistics vary over time. The variations are referred to as either **climate variability** or **climate change**. Climate variability is the variation of climate on a seasonal or inter-annual basis. A "warmer-than-normal" summer followed by a "cooler-than-normal" autumn is considered climate variability. Climate change describes measurable shifts in the climate statistics over an extended period, such as a century or longer. Past glacial ice ages are examples of climate changes.

The most commonly used statistics to describe the climate include the **mean, standard deviation, median, mode, normal, range,** and **trend**. These particular statistics are defined and illustrated below using the following **sequence** of 10 numbers:

$$[1, 4, 8, 3, 6, 4, 6, 9, 4, 7]$$

Mean

The mean is the arithmetic average of a **series** of data, obtained by summing all of the values and dividing by the number of observations. Mathematically, the mean is defined as follows:

$$\overline{X} = \frac{\sum X_i}{N},$$

where $\overline{X}$ is the mean, X_i are the individual values of the series, and N is the number of observations. (Note: In statistics, the Greek letter $\sum$ represents the summation of a series. Hence, in this case, $\sum X_i = 1 + 4 + 8 + 3 + 6 + 4 + 6 + 9 + 4 + 7$.)

The mean of the example dataset is equal to

$$\overline{X} = (1 + 4 + 8 + 3 + 6 + 4 + 6 + 9 + 4 + 7)/10 = 5.2.$$

Standard Deviation

The standard deviation is a measure of the variation, or spread, of the observations about their mean. Mathematically, the standard deviation is defined as follows:

$$\sigma = \sqrt{\frac{\sum (X_i - \overline{X})^2}{N - 1}},$$

where σ is the standard deviation, $\overline{X}$ is the mean, X_i are the individual values of the series, and N is the number of observations.

The standard deviation of the example dataset is equal to

$$\sigma = \sqrt{\frac{(1 - 5.2)^2 + (4 - 5.2)^2 + ... + (7 - 5.2)^2}{10 - 1}} = \sqrt{\frac{53.6}{9}} = 2.44.$$

Median

The median is the middle value of a ranked data series, sorted either from highest to lowest or vice-versa. The median is the midpoint of the series, determined by:

$$Median = X_j \ ,$$

where $j = \dfrac{(N+1)}{2}$ in a ranked series, and X_j is the individual value of the ranked series for item j.

In the case of an even number of data points in the series, the median is determined by averaging the values immediately before and after the index value.

In the example dataset, the ranked series is [1, 3, 4, 4, 4, 6, 6, 7, 8, 9]. The midpoint is the value for item

$$j = \dfrac{(10+1)}{2} = 5.5 \ .$$

So, in this case, the median value is the average of the 5th and 6th values of the series. Hence, because $X_5 = 4$ and $X_6 = 6$, the median is (4+6)/2, or 5.

The median is a useful measure in cases when an extreme value shifts the mean such that it is not representative of the series. For example, if the value "9" in our sample series were replaced with "99," the mean would increase from 5.2 to 14.2 — a value greater than all but one of the numbers in the data series. Despite this new extreme value, the median would remain 5.

Mode

The mode is the value that occurs most frequently in the series. In the example dataset, the number "4" appears three times, the number "6" appears twice, and all other numbers each appear only once. Therefore, the mode of the sample series is 4.

The mode provides an idea of the uniformity of a series. For example, if there is a relatively steady "base state" with a few data anomalies superimposed, the mode reveals that base state. In a sample series [0, 0, 0, 0, 1, 0, 0, 3, 0, 0], the value "0" appears eight times and the values "1" and "3" each appear only once. In this instance, the "1" and "3" are anomalies superimposed on the base state, or mode, of 0 (zero).

A given dataset may have more than one mode. For example, in the sample series [0, 1, 2, 2, 3, 5, 5, 7, 9, 9], the values "2," "5," and "9" each occur twice. Thus, the series has three modes.

Normal

Perhaps the most common description of climate for a given day or month is the "normal temperature" or "normal precipitation." Technically, the normal value is determined by fitting a **Gaussian** distribution to a data series (e.g., of daily maximum temperatures for 30 years). An example of a Gaussian distribution is the "bell curve." The midpoint of the curve is the "normal" value of the series. In most instances, the normal is nearly identical to the mean of the same series.

Climatologists typically use a 30-year reference period to determine daily and monthly normals because it is long enough such that the mean does not fluctuate widely. They update this 30-year window every ten years. For example, between 1991 and 2000, the "normal" high temperature for a given calendar day (e.g., June 22) was based on the average high temperature of the 30 such observations (e.g., on all June 22nds) during the period 1961-1990. During the following decade, normals were established using the 1971-2000 period as the reference. Because it uses only a subset of the overall climate record, the normal rarely equals the mean of the entire record. Because the normals are updated every 10 years, one must interpret climate variability carefully; what may appear to be a shift in climate patterns may be simply a change in the normals when comparing two distinct time periods.

Range

The range is the lowest value of the series subtracted from the highest value. If the series is ranked from lowest to highest, then the range can be calculated as follows:

$$Range = X_N - X_1 ,$$

where X_N is the highest value and X_1 is the lowest value of the ranked series.

The range of the example dataset is equal to

$$Range = X_{10} - X_1 = 9 - 1 = 8.$$

The range is useful for determining extreme values of the distribution.

Trend and Trendline

A **trendline** displays a trend computed from a sequence of data, preferably with many data values. Trendlines can be calculated using **linear regression** to fit a straight line to a sequence of data. The straight line has a slope, m, and an intercept, b, where it crosses y-axis (e.g., $X = 0$). In climatology, the values of X typically describe a time period and represent years, months, or days. Mathematically, this line is represented by the following equation:

$$Y = mX + b.$$

The slope, m, is defined as the **covariance** divided by the **variance**:

$$Covariance(XY) = \sum (X_i - \overline{X})(Y_i - \overline{Y})$$

and

$$Variance(X) = \sum (X_i - \overline{X})^2 ,$$

where $\overline{X}$ and $\overline{Y}$ are the means, and X_i and Y_i are the individual values of the series.
(Note: Variance is identical to the square of the standard deviation.)

The intercept, b, is calculated from the means of the series as follows:

$$b = \overline{Y} - m\overline{X}.$$

Consider the sample sequence [1, 4, 8, 3, 6, 4, 6, 9, 4, 7] for the variable Y, and a year from 1 to 10 for the variable X. For example, Y might represent average minimum temperature for January and X might represent the period from 1991 to 2000. Table 1 demonstrates how to use the equations above to calculate the trendline.

Table 1 – Example Dataset for Calculation of a Trendline

	X (Year)	Y (Observation)	$X_i - \overline{X}$	$(X_i - \overline{X})^2$	$Y_i - \overline{Y}$	$(X_i - \overline{X})(Y_i - \overline{Y})$
	1	1	-4.5	20.25	-4.2	18.9
	2	4	-3.5	12.25	-1.2	4.2
	3	8	-2.5	6.25	2.8	-7
	4	3	-1.5	2.25	-2.2	3.3
	5	6	-0.5	0.25	0.8	-0.4
	6	4	0.5	0.25	-1.2	-0.6
	7	6	1.5	2.25	0.8	1.2
	8	9	2.5	6.25	3.8	9.5
	9	4	3.5	12.25	-1.2	-4.2
	10	7	4.5	20.25	1.8	8.1
Sum	55	52		82.5		33
Average	$\overline{X}$ =5.5	$\overline{Y}$ = 5.2				

Using the results from Table 1, the slope and intercept of the trendline are calculated as follows:

$$m = \frac{covariance\,(XY)}{variance\,(X)} = \frac{\sum (X_i - \overline{X})(Y_i - \overline{Y})}{\sum (X_i - \overline{X})^2} = \frac{33}{82.5} = 0.4$$

and

$$b = \overline{Y} - m\overline{X} = 5.2 - (0.4)(5.5) = 3.0 \ .$$

Hence, any point along the trendline can be computed from the year, X, using the following equation:

$$Y = mX + b = 0.4X + 3.0 \ .$$

Using this equation, you can determine that the first data value on the trendline (at X = 1) would be 3.4 and the last value (at X = 10) would be 7.0 — an increase of 3.6 over the span of the sequence. The sign (either positive or negative) of the slope m represents the trend. For example, a warming trend has a positive slope and a cooling trend has a negative slope. If the example dataset were temperature, the slope (m = 0.4) would represent warming over time.

FREQUENCY DISTRIBUTION TOOLS

Climate also can be characterized using tools such as **frequency distributions** or **probability distributions**. Frequency distributions categorize observations into discrete intervals and portray accumulated values using a chart or graph. Figure 1 shows the distribution of maximum temperatures during 2003 at Freedom, OK. The x-axis partitions the observed temperatures into 10°F bins, and the y-axis represents the number of days during 2003 when the observed maximum temperature fell within each of the bins.

Charts that display frequency distributions (such as Figure 1) show the actual distribution of data and are useful to visualize the most and least commonly observed values. However, this type of visualization does not answer the often-asked question: "How likely will a certain temperature occur?" For example, what is the probability that the temperature was between 70 and 79 degrees at Freedom, OK? To answer this question, the number of occurrences of each category (e.g., on the y-axis in Figure 1) is converted to a percentage by dividing the number for each category by the total number of observations. These percentages then can be plotted as a continuous function, known as a **probability distribution function**. Probability distribution functions represent the probability that a variable will be a specific value, *V*. Figure 2 displays a probability distribution function for the data originally plotted in Figure 1. In this case, the **probability of occurrence** of a maximum temperature at Freedom between 70°F and 79°F during 2003 was about 20%. Similarly, the probability of Freedom's maximum temperature being between 30°F and 39°F was about 6%. More elaborate statistical techniques may be used to fit the data into different statistical distributions.

Often, climatologists need to provide decision makers with information about the probability that a value will exceed some threshold. To compute this **probability of exceedence**, the climatologist first calculates a **cumulative distribution function**. A cumulative distribution function describes the frequency that a variable has a value, *V*, less than or equal to a given threshold, *T*. As demonstrated by Figure 3, at small values of *T*, the frequency of that value or less occurring is near zero; at large values of *T*, the frequency of that value or less occurring is near 100%. In other words, at Freedom during 2003, maximum temperatures less than 40°F occurred ~10% of the time. Ninety percent of the time the maximum temperatures were less than 100°F. And the 100% probability of occurrence at 120°F means that *every* observation at Freedom was less than or equal to 120°F.

The climatologist then calculates the probability of exceedence by inverting the graph, changing the likelihood of values *less than and equal to T* to values *greater than T*. Mathematically, the probability of exceedence is equal to 100% minus the frequency (probalility) of occurrence for each value of *V*. As demonstrated by Figure 4, for small values of *T*, the probability of exceedence is approximately 100%; at large values of *T*, the probability of exceedence is approximately 0%. For our example from Freedom,

the probability of a temperature *between* 70°F and 79°F was 20% (Figure 2), the probability of a temperature *less than* 70°F was about 40% (Figure 3), and the probability of a temperature *exceeding* 70°F is about 60% (Figure 4).

Cumulative distribution functions can be extremely useful to estimate the likelihood of an extreme event. Extreme events can be described using the cumulative distribution function through the concept of a **return period**. A return period is the average interval (usually in years) between events of a specified magnitude. For example, a daily rainfall total of 3 inches may happen once every five years on average, but a daily rainfall total of 4 inches may occur only once every 25 years. An event occurring once every five years would occur, on average, 20 times during a 100-year sample, or 20% of the time; a "25-year event" would happen only 4 times, or 4%. These numbers result from the probability of exceedence.

Return periods actually are statements of probability. Hence, because they are based on statistics, there is no guarantee that after a "25-year event" has occurred, it will not happen the following year. It only means that using the existing 100-year record, there should be four such rainfall events (i.e., an average of one event every 25 years). These events could occur in consecutive years or even multiple times during a single year!

Decision-makers can use return periods as a first-guess of the severity of a given *weather event*. For example, if a daily rainfall total of 6 inches occurs once every 100 years (i.e., the 100-year flood event), the decision-maker who monitors rainfall totals in real-time can recognize when a given event passes the 6-inch threshold. Thus, the decision-maker can plan accordingly to deal with the consequences of a rare event.

Figure 1 – Frequency Distribution of Maximum Temperature for Freedom, OK during 2003

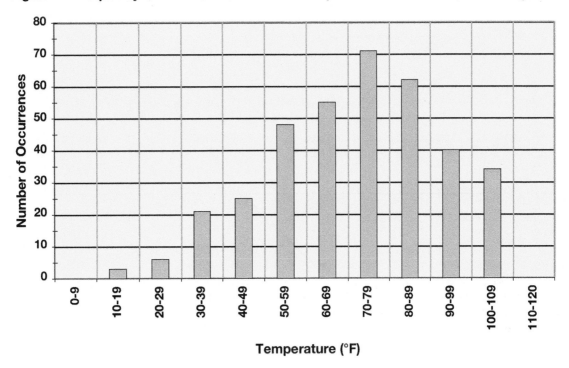

Figure 2 – Probability Distribution Function of Maximum Temperature at Freedom, OK for 2003

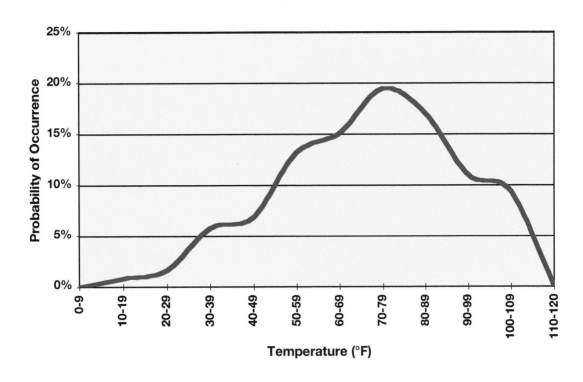

Figure 3 – Cumulative Distribution Function of Maximum Temperature at Freedom, OK for 2003

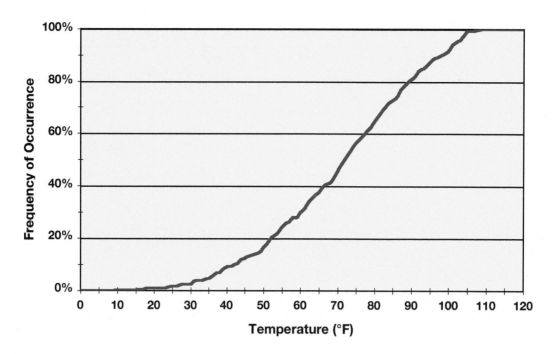

Figure 4 – Probability of Exceedence for Maximum Temperature at Freedom, OK during 2003

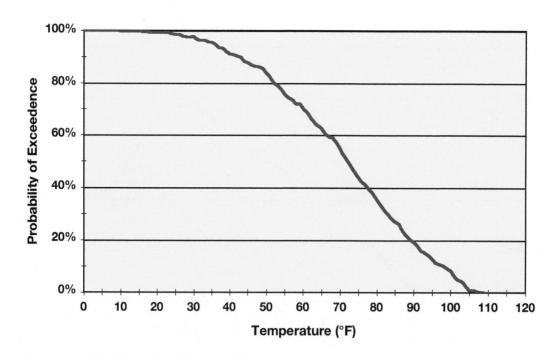

LABORATORY EXERCISES:

Part I: Weather and Climate

Table 2 lists the maximum and minimum temperatures (in °F) and precipitation (in inches) for every day during April 2003 at Fairbanks, Alaska. You will compare your calculations of several climate statistics with the daily values.

1. Calculate the mean, median, mode, and standard deviation of the daily data for each of the variables in Table 2.

Maximum Temperature	**Minimum Temperature**	**Precipitation**
Mean _____ °F	Mean _____ °F	Mean _____ in
Median _____ °F	Median _____ °F	Median _____ in
Mode _____ °F	Mode _____ °F	Mode _____ in
Std. Dev. _____ °F	Std. Dev. _____ °F	Std. Dev. _____ in

Table 2 – Daily Maximum and Minimum Temperatures and Daily Precipitation for Fairbanks, AK, during April 2003

Day	Maximum Temperature (°F)	Minimum Temperature (°F)	Precipitation (inches)
1	16	−14	0
2	23	−12	0
3	26	−4	0
4	31	−3	0
5	31	3	0
6	28	12	0
7	23	8	0
8	36	3	0
9	44	11	0
10	50	19	0
11	53	21	0
12	54	24	0
13	42	34	0.01
14	42	27	0.01
15	45	27	0.01
16	45	29	0.01
17	48	32	0.04
18	52	30	0
19	48	29	0
20	52	27	0
21	54	30	0
22	58	32	0
23	58	36	0
24	65	31	0
25	62	31	0
26	68	33	0
27	67	34	0
28	48	32	0.01
29	50	29	0
30	50	27	0

2. How many times did the daily observations in Table 2 match the *means* that you calculated in question 1?

Max. Temp. _____ Min. Temp. _____ Precip. _____

3. How many times did the daily observations in Table 2 match the *modes* that you calculated in question 1?

Max. Temp. _____ Min. Temp. _____ Precip. _____

4. How many times did the daily observations in Table 2 lie within *one standard deviation* of the means that you calculated in question 1?

Max. Temp. _____ Min. Temp. _____ Precip. _____

5. Using your answers from questions 2 to 4, discuss how well the various types of climate statistics characterize the weather of April 2003 in Fairbanks, as represented by the data in Table 2.

_____

_____

_____

_____

_____

_____

_____

Part II: Climate Variability

Table 3 lists the annual average temperatures (in °F) and annual average precipitation (in inches) from 1881 to 2000 for Philadelphia, Pennsylvania. Table 4 presents the mean, median, and standard deviation of the average temperatures for the periods of 1881–2000 and 1971–2000. To examine the variability of the climate, you will compare climate statistics for the period from 1881 to 2000 with those for the period from 1971 to 2000.

6. Using Tables 3 and 4, describe how the 30-year average differs from the long-term average? What accounts for these differences?

_____

_____

_____

_____

Table 3 – Annual Average Temperature and Precipitation
for Philadelphia, PA from 1881 to 2000

Year	Annual Average Temp. (°F)	Annual Average Precip. (inches)	Year	Annual Average Temp. (°F)	Annual Average Precip. (inches)	Year	Annual Average Temp. (°F)	Annual Average Precip. (inches)	Year	Annual Average Temp. (°F)	Annual Average Precip. (inches)
1881	54.5	29.91	1911	55.4	51.35	1941	54.8	35.15	1971	55.6	47.79
1882	54.8	45.58	1912	54.2	47.00	1942	54.3	43.05	1972	54.0	49.63
1883	54.1	39.18	1913	57.0	47.41	1943	53.7	36.77	1973	56.3	46.06
1884	54.2	39.34	1914	54.3	39.07	1944	54.2	39.52	1974	55.3	37.78
1885	51.9	33.35	1915	55.4	44.83	1945	54.6	46.68	1975	56.0	52.13
1886	53.7	37.24	1916	54.3	32.27	1946	55.4	38.35	1976	54.2	33.27
1887	54.6	42.17	1917	52.7	39.39	1947	54.1	44.46	1977	54.3	49.42
1888	52.9	44.06	1918	54.8	37.73	1948	54.1	49.07	1978	53.4	45.95
1889	54.8	50.60	1919	55.7	49.12	1949	57.0	40.48	1979	54.4	52.79
1890	55.0	34.02	1920	54.4	46.16	1950	54.3	40.47	1980	54.4	38.80
1891	55.1	38.19	1921	57.1	35.45	1951	55.4	41.75	1981	53.7	37.83
1892	53.5	34.78	1922	55.7	29.31	1952	55.9	45.84	1982	54.1	40.43
1893	53.0	37.65	1923	55.3	39.19	1953	56.8	48.13	1983	54.7	54.41
1894	55.0	40.34	1924	53.8	43.11	1954	55.7	34.04	1984	53.7	43.66
1895	53.7	31.01	1925	55.9	32.40	1955	55.4	33.03	1985	54.8	35.20
1896	54.5	32.15	1926	53.6	44.91	1956	54.2	46.00	1986	55.3	40.42
1897	54.6	42.04	1927	55.5	43.15	1957	55.1	32.20	1987	55.4	33.40
1898	55.6	49.23	1928	55.4	39.37	1958	52.6	47.87	1988	54.5	38.41
1899	54.5	39.96	1929	55.9	41.56	1959	55.5	38.37	1989	54.4	48.66
1900	56.1	40.91	1930	56.8	33.97	1960	52.8	41.15	1990	57.5	35.79
1901	53.8	45.54	1931	58.1	39.28	1961	52.7	41.05	1991	58.0	36.22
1902	54.2	49.76	1932	57.1	44.52	1962	52.1	42.62	1992	55.2	30.41
1903	54.2	41.50	1933	56.4	51.37	1963	51.9	34.95	1993	56.7	42.18
1904	51.8	39.76	1934	55.1	38.36	1964	54.0	29.88	1994	56.6	44.92
1905	53.9	41.61	1935	54.4	46.36	1965	53.0	29.34	1995	56.5	31.53
1906	55.4	51.87	1936	54.9	38.70	1966	53.0	40.00	1996	53.7	56.47
1907	53.1	48.74	1937	55.3	37.40	1967	53.3	44.82	1997	54.7	32.52
1908	55.5	38.13	1938	56.0	46.92	1968	54.1	35.45	1998	58.1	31.66
1909	55.1	37.36	1939	55.8	45.40	1969	53.8	43.36	1999	56.7	48.50
1910	55.0	39.60	1940	52.1	44.61	1970	54.4	39.14	2000	54.6	44.20

Table 4 – Mean, Median, and Standard Deviation for the Variables in Table 2
for the Periods 1881–2000 and 1971–2000

	Annual Average Temp. (°F)	Annual Average Precip. (inches)
Mean (1881 – 2000)	54.8	41.1
Median (1881 – 2000)	54.6	40.5
Std. Dev. (1881 – 2000)	1.3	6.1
Mean (1971 – 2000)	55.2	42.0
Median (1971 – 2000)	54.8	41.3
Std. Dev. (1971 – 2000)	1.3	7.5

7. Calculate a trendline for annual average temperature from 1971 to 2000 using the data in Table 5. To compute the trendline, fill in the blank boxes that are highlighted in Table 5. Then use your calculated variance and covariance to complete the equations on the next page.

Table 5 – Values of Annual Average Temperature for Philadelphia, PA from 1971 to 2000 Used to Calculate a Trendline

	X	Y	$X_i - \overline{X}$	$(X_i - \overline{X})^2$	$Y_i - \overline{Y}$	$(X_i - \overline{X})(Y_i - \overline{Y})$
	1	55.6	-14.5	210.25	0.4	-5.80
	2	54.0	-13.5	182.25	-1.2	16.20
	3	56.3	-12.5	156.25	1.1	-13.75
	4	55.3	-11.5	132.25	0.1	-1.15
	5	56.0	-10.5	110.25	0.8	-8.40
	6	54.2				
	7	54.3	-8.5	72.25	-0.9	7.65
	8	53.4	-7.5	56.25	-1.8	13.50
	9	54.4	-6.5	42.25	-0.8	5.20
	10	54.4	-5.5	30.25	-0.8	4.40
	11	53.7	-4.5	20.25	-1.5	6.75
	12	54.1	-3.5	12.25	-1.1	3.85
	13	54.7				
	14	53.7	-1.5	2.25	-1.5	2.25
	15	54.8	-0.5	0.25	-0.4	0.20
	16	55.3				
	17	55.4	1.5	2.25	0.2	0.30
	18	54.5	2.5	6.25	-0.7	-1.75
	19	54.4	3.5	12.25	-0.8	-2.80
	20	57.5	4.5	20.25	2.3	10.35
	21	58.0	5.5	30.25	2.8	15.40
	22	55.2				
	23	56.7	7.5	56.25	1.5	11.25
	24	56.6	8.5	72.25	1.4	11.90
	25	56.5	9.5	90.25	1.3	12.35
	26	53.7	10.5	110.25	-1.5	-15.75
	27	54.7	11.5	132.25	-0.5	-5.75
	28	58.1				
	29	56.7	13.5	182.25	1.5	20.25
	30	54.6	14.5	210.25	-0.6	-8.70
Sum	465	1656.8				
Average	$\overline{X} = 15.5$	$\overline{Y} = 55.2$				

$$m = \frac{\sum (X_i - \overline{X})(Y_i - \overline{Y})}{\sum (X_i - \overline{X})^2} = \frac{\underline{\hspace{2cm}}}{\underline{\hspace{2cm}}} = \underline{\hspace{2cm}}$$

$$b = \overline{Y} - m\overline{X} = \underline{\hspace{2cm}} - (\underline{\hspace{2cm}})(\underline{\hspace{2cm}}) = \underline{\hspace{2cm}}$$

The final equation for the trendline is _____

8. Examine your results from question 7. Did the climate change during the 30-year period at the Philadelphia station? How does the trend compare to the variability measured by the 1971–2000 standard deviation (Table 4)?

_____

_____

_____

_____

_____

_____

_____

9. Figures 5 and 6 show the distributions of annual average temperature and annual average precipitation, respectively, for Philadelphia, PA from 1881 to 2000. What factors account for differences in the shapes of the distributions?

_____

_____

_____

_____

_____

_____

_____

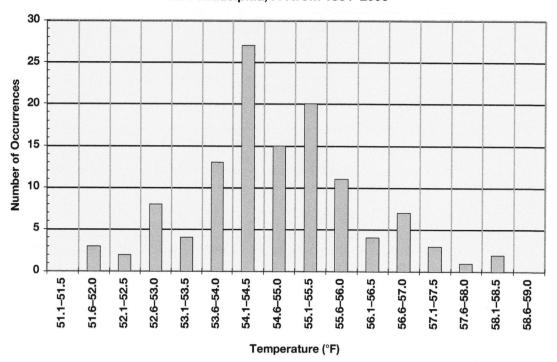

Figure 5 – Frequency Distribution of Average Annual Temperature
for Philadelphia, PA from 1881–2000

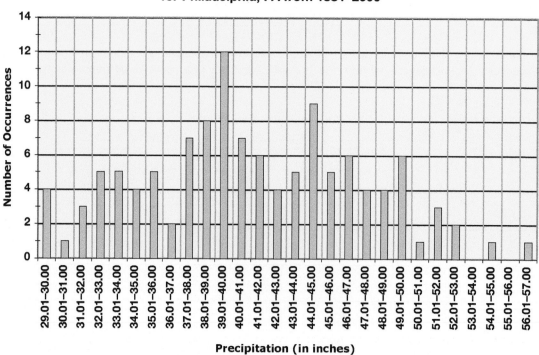

Figure 6 – Frequency Distribution of Average Annual Precipitation
for Philadelphia, PA from 1881–2000

10. Compare trendlines on the graphs of annual average temperature from downtown Baltimore, MD (Figure 7), and from Woodstock, MD (Figure 8), located near but outside the Baltimore metropolitan area. What difference, if any, exists between the intercepts on these graphs? Also, compare the long-term trends of the two sites. What conclusions can you draw about climate trends in the Baltimore–Woodstock area based on the 100-year datasets? Speculate on any causes of the climate trends.

_____

_____

_____

_____

_____

Figure 7 – Annual Average Temperature (in °F) for Baltimore, MD from 1901 to 2000

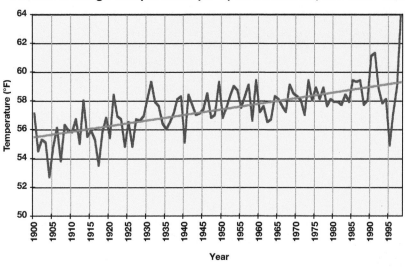

Figure 8 – Annual Average Temperature (in °F) at Woodstock, MD from 1901 to 2000

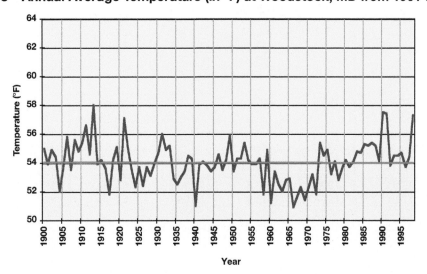

11. Compare the annual means and standard deviations for the cities listed in Table 6. What two cities have the means closest to each other? What two cities have the greatest variation in their means? Describe the factors that might cause El Paso, TX, to have greater climate variability than Key West, FL.

Two cities with closest means _____

Two cities with greatest variance in their means _____

_____

_____

_____

_____

Table 6 – Annual Means and Standard Deviations for Four Selected Cities
(Key West, FL; Spokane, WA; El Paso, TX; Burlington, VT)

Key West, FL		Spokane, WA		El Paso, TX		Burlington, VT	
Year	Annual Average Temperature (in °F)	Year	Annual Average Temperature (in °F)	Year	Annual Average Temperature (in °F)	Year	Annual Average Temperature (in °F)
1971	78.13	1971	47.07	1971	63.25	1971	43.47
1972	78.85	1972	46.76	1972	63.83	1972	42.36
1973	77.78	1973	48.49	1973	61.91	1973	46.12
1974	78.60	1974	47.67	1974	62.69	1974	44.28
1975	79.31	1975	45.42	1975	62.51	1975	46.04
1976	77.49	1976	47.07	1976	61.37	1976	43.72
1977	76.84	1977	46.90	1977	63.76	1977	45.33
1978	77.55	1978	45.73	1978	64.73	1978	42.72
1979	78.33	1979	46.91	1979	62.38	1979	45.40
1980	77.94	1980	47.31	1980	64.34	1980	44.27
1981	77.06	1981	47.84	1981	65.03	1981	45.91
1982	78.98	1982	47.07	1982	64.13	1982	45.20
1983	76.62	1983	48.53	1983	63.17	1983	45.78
1984	77.28	1984	46.29	1984	63.06	1984	45.62
1985	77.98	1985	43.74	1985	62.19	1985	44.50
1986	78.23	1986	47.70	1986	63.16	1986	44.95
1987	77.45	1987	48.83	1987	61.40	1987	45.49
1988	77.34	1988	48.08	1988	62.37	1988	45.47
1989	78.62	1989	46.93	1989	64.11	1989	44.61
1990	79.30	1990	48.09	1990	63.99	1990	47.54
1991	79.43	1991	47.47	1991	62.98	1991	47.03
1992	77.99	1992	49.45	1992	64.64	1992	44.30
1993	78.22	1993	45.33	1993	65.74	1993	44.67
1994	79.03	1994	49.03	1994	67.39	1994	45.04
1995	78.18	1995	48.03	1995	65.92	1995	46.82
1996	77.01	1996	45.36	1996	65.46	1996	45.28
1997	78.48	1997	47.91	1997	63.91	1997	44.39
1998	78.41	1998	50.33	1998	64.59	1998	48.28
1999	77.94	1999	48.20	1999	64.89	1999	47.84
2000	77.64	2000	45.98	2000	65.57	2000	44.94
Mean	78.07		47.32		63.82		45.25
Std. Dev.	0.75		1.38		1.43		1.37

Part III: Return Periods

12. **(Advanced Students/Meteorology Majors)** If a 25-year rainfall event occurs this year, how likely is a similar event to occur next year as compared to 20 years from now? Briefly explain your answer.

_____

_____

_____

_____

13. **(Advanced Students/Meteorology Majors)** Using the dataset of annual precipitation for Savannah, GA in Table 7 and the probablility of exceedence graph in Figure 9, calculate the value of precipitation for return periods of 5, 10, 25, and 100 years. That is, what is the maximum amount of precipitation that Savannah could expect during a 5-year period, 10-year period, etc.?

Max. Precip. for a 5-year Return Period is _____ inches

Max. Precip. for a 10-year Return Period is _____ inches

Max. Precip. for a 25-year Return Period is _____ inches

Max. Precip. for a 100-year Return Period is _____ inches

Table 7 – Annual Precipitation Recorded for Savannah, GA
for the 100–Year Period 1901-2000

Year	Annual Precipitation (in inches)	Year	Annual Precipitation (in inches)	Year	Annual Precipitation (in inches)	Year	Annual Precipitation (in inches)
1901	36.84	1926	54.66	1951	48.66	1976	63.74
1902	47.35	1927	35.18	1952	38.47	1977	41.84
1903	53.67	1928	55.67	1953	55.25	1978	35.41
1904	40.42	1929	53.90	1954	32.83	1979	61.92
1905	45.49	1930	42.15	1955	38.52	1980	37.84
1906	39.67	1931	22.00	1956	43.16	1981	40.06
1907	42.45	1932	46.78	1957	63.96	1982	52.26
1908	47.56	1933	48.63	1958	48.01	1983	54.51
1909	38.30	1934	42.34	1959	68.00	1984	50.66
1910	46.61	1935	43.89	1960	53.73	1985	38.64
1911	36.17	1936	39.75	1961	49.00	1986	45.33
1912	61.84	1937	51.47	1962	54.08	1987	56.70
1913	40.00	1938	29.81	1963	50.83	1988	48.17
1914	43.58	1939	35.96	1964	73.17	1989	46.87
1915	52.15	1940	36.54	1965	45.81	1990	43.08
1916	37.91	1941	45.30	1966	45.39	1991	68.42
1917	42.15	1942	38.69	1967	41.27	1992	58.36
1918	43.14	1943	40.83	1968	37.34	1993	48.05
1919	48.23	1944	61.01	1969	60.84	1994	69.44
1920	57.84	1945	55.91	1970	53.84	1995	51.11
1921	42.01	1946	43.14	1971	61.34	1996	36.16
1922	58.26	1947	67.25	1972	48.57	1997	54.42
1923	32.62	1948	64.34	1973	45.40	1998	49.47
1924	57.72	1949	41.16	1974	41.93	1999	48.78
1925	41.69	1950	55.94	1975	51.18	2000	49.85

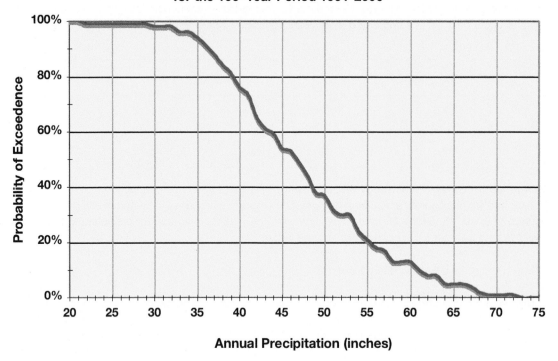

Figure 9 – Probability of Exceedence for Annual Precipitation for Savannah, GA for the 100–Year Period 1901-2000

Annual Precipitation (inches)

14. **(Advanced Students/Meteorology Majors)** Using Table 7, how often does the annual precipitation for Savannah exceed 60 inches? What is the least and greatest number of years between occurrences of this threshold?

The annual precipitation exceeds 60 inches every _____ years.

Least number of years between 60-inch events _____

Greatest number of years between 60-inch events _____

GLOSSARY

A

Absorption – The process of capturing and retaining incident radiant energy in a substance.

Adiabat – On a thermodynamic diagram, a line representing the temperature change with height for either a dry (unsaturated) adiabatic process (i.e., a dry adiabat) or a moist (saturated) adiabatic process (i.e., a moist adiabat).

Adiabatic – Any thermodynamic process whereby no heat is exchanged between an object (e.g., an air parcel) and its surrounding environment.

Advection – The transport of an atmospheric property (e.g., temperature) by the wind.

Air Mass – A body of air that extends hundreds or thousands of kilometers horizontally and has relatively uniform temperature and moisture characteristics (e.g., warm/moist, cool/dry).

Air Parcel – An imaginary body of air a few meters in diameter that possesses nearly uniform properties within it.

Albedo – The fraction of incident radiation that reflects off a body.

Altimeter Setting – A measure of atmospheric pressure used to indicate the altitude with respect to a fixed level above sea level; used by pilots.

Angle of Incidence – The angle at which a ray of light (or radiation) strikes a surface. In meteorology, it is measured as the angle between an incoming ray of sunlight and a line perpendicular to the earth's surface at the point of incidence (i.e., where the ray strikes).

Asymmetric – Not symmetric.

Atmospheric Instability – The atmospheric state in which an air parcel that is pushed slightly either upward or downward will continue to move in the direction it was pushed. The condition exists when a warm, moist air parcel (i.e., buoyant parcel) rises with little external influence and remains warmer than the surrounding air as it rises.

Autumnal Equinox – The equinox when the sun has reached a point directly above the equator on its southward journey into the Southern Hemisphere. It marks the start of the astronomical autumn (~21 September) in the Northern Hemisphere. On that day, daylight lasts 12 hours everywhere on the earth and the sun is directly over the equator at solar noon.

B

Base Reflectivity – A radar product from a Doppler radar that indicates the location and intensity of precipitation. Base reflectivity is associated with the size, number, and type of particles (within the radar beam) that are capable of scattering the radar's emitted energy back to the radar antenna. Hence, base reflectivity is related to rainfall intensity (e.g., drop size and rainfall rate) and hail size (for large values of reflectivity).

Blackbody – A hypothetical "body" that absorbs all of the electromagnetic radiation striking it. A blackbody does not reflect or transmit any of the incident radiation. A blackbody not only absorbs energy at all wavelengths, but also emits energy at all wavelengths with the maximum possible intensity for any given temperature.

Buoyant – Tending to float or to rise when submerged in a liquid or gas. A buoyant object is less dense than the surrounding fluid.

C

Cirrus – A high-level cloud (~5 km or higher) that is composed mostly of ice crystals and has the appearance of white, delicate filaments in patches or narrow bands.

Clear–Air Mode – A highly sensitive operational mode of NEXRAD Doppler radar in which the antenna scans slowly to sense echoes from "clear-air" (i.e., no precipitation). In clear-air mode, the radar obtains 19 elevation slices in 10 minutes. These echoes can be from dirt, insects, smoke, and changes in the air density. Clear-Air mode is used widely during snow events.

Climate – The statistical collective summary of weather conditions at a place over a period of years.

Climate Change – A significant, long-term change in the climatic state of a locale or a large area (i.e., a change resulting in important economic, environmental, or social effects).

Climate Variability – Deviations of climate statistics during a given period of time (e.g., a specific month, season, or year) from the long-term climate statistics relating to the corresponding calendar interval.

Cloud-to-Ground Lightning – A lightning discharge between a thunderstorm and the ground.

Cold Front – The leading edge of an advancing cold air mass.

Composite Chart – A map constructed by overlaying critical values of atmospheric ingredients. A meteorologist uses a composite chart to assess the potential for significant weather. A composite chart may indicate the relative positions of regions of low-level moisture, a thermal ridge, a 300-mb jet stream, and a 500-mb height trough.

Condensation – The (isothermal) process by which a gas (e.g., water vapor) changes into a liquid.

Conduction – The transfer of energy by molecular motion from warmer to colder regions through a substance or between objects in direct contact, and without any net external motion.

Contiguous – Touching or connected throughout an unbroken sequence.

Continental Arctic Air Mass – An air mass characterized by extremely cold, dry air.

Continental Polar Air Mass – An air mass characterized by cold, dry air.

Continental Tropical Air Mass – An air mass characterized by warm or hot, dry air.

Contour Line – A line of constant value of some variable (e.g., height). Also known as an isopleth. Contour lines separate larger values of a quantity from smaller values on a map.

Convection – In general, the transport and mixing of the properties of a fluid (e.g., heat, moisture, etc.) by means of mass motion within the fluid. In meteorology, atmospheric motions generally are divided into those in the horizontal, or advection, and those in the vertical, or convection. Convection typically results from surface heating and the subsequent rising of warm air.

Convective Condensation Level (CCL) – The level in the atmosphere to which an air parcel, if sufficiently heated from below, will rise dry adiabatically until it first becomes saturated and becomes as warm as the environmental temperature. On a Skew-T diagram, the CCL is located at the intersection of the environmental temperature curve and the mixing ratio line through the surface dew point.

Convective Precipitation – Precipitation resulting from a cloud that has developed vertically by convection.

Convergence – The net inflow of air into an area. When convergence occurs at the surface, rising motion results.

Coriolis Force – An apparent force that, as a result of the earth's rotation, deflects objects moving above the earth's surface to the right in the Northern Hemisphere and to the left in the Southern Hemisphere.

Covariance – A measure of the extent to which two variables (e.g., month and average temperature) are related to one another.

Cumulative Distribution Function – A function that represents the likelihood that an observation will be less than or equal to a selected value.

Cumulonimbus – An exceptionally dense and vertically developed cloud type, occurring both as isolated clouds and as a line or wall of clouds, and generally accompanied by heavy rain, lightning, and thunder.

Cumulus – Cloud type in the form of individual, detached elements which are generally dense, have well-defined outlines, show vertical development in the form of domes, mounds, or towers.

Cumulus Stage (of a Thunderstorm) – The stage in the life cycle of a typical thunderstorm that is characterized by a large cumulus cloud composed mainly of an ascending column of air. In this stage, which may last for 10-15 minutes, little or no rain falls.

D

Daily Temperature Cycle – The change in air temperature during a given 24-hour period, usually from midnight to midnight. Also known as the diurnal temperature cycle.

Dalton's Law of Partial Pressures – For ideal gases, a mixture of gases will have a pressure equal to the sum of the pressures of the individual gases, assuming no chemical reaction has taken place between the gases. Named for John Dalton (1766 – 1844), a British chemist who formulated the concept.

Developing Stage (of a Thunderstorm) – See Cumulus Stage (of a Thunderstorm).

Dew Point (or Dewpoint Temperature) – A measure of atmospheric moisture. The dew point is the temperature to which air must be cooled for saturation to occur (given a constant pressure and constant water vapor content).

Dewpoint Depression – The difference in degrees between the air temperature and the dewpoint temperature. Dewpoint depression is a measure of atmospheric moisture content.

Dissipating Stage (of a Thunderstorm) – The stage in the life cycle of a typical thunderstorm that occurs when the storm is dominated by downdrafts, cutting off the supply of warm, moist, unstable air that feeds the storm's updraft.

Diurnal – Daily; related to actions which are completed during a single calendar day, and which typically recur every calendar day (e.g., diurnal temperature cycle of temperature increase and decrease).

Divergence – The net outflow of air from an area. Divergence at the surface is accompanied by sinking motion.

Downburst – An intense localized downdraft that may be experienced beneath a thunderstorm, typically a severe thunderstorm; it results in an outward burst of damaging winds on or near the ground.

Downdraft – A relatively small-scale current of air with marked downward motion.

Downwelling Radiation – The component of radiation directed toward the earth's surface from the sun or the atmosphere.

Dry Adiabat – On a thermodynamic diagram, a line representing the temperature change with height for a dry (unsaturated) adiabatic process.

Dry Adiabatic Lapse Rate – The rate at which air temperature changes with height for an unsaturated parcel of air.

Dry-Bulb Temperature – Same as air temperature.

Dryline – A boundary separating warm, dry air from warm, moist air. In the central United States, a dryline typically lies north-to-south across the central and southern high Plains states during the spring and early summer. It separates moist air from the Gulf of Mexico (to the east) and dry desert air from the southwestern states (to the west).

E

Electromagnetic Spectrum – The ordered series of all known types of electromagnetic radiation, arranged by wavelength from cosmic rays (very short wavelength) through gamma rays, X-rays, ultraviolet radiation, visible radiation, infrared radiation, microwaves, to AM and FM radio and television (long wavelengths).

Electromagnetic Waves – The waves from an advancing disturbance in electric and magnetic fields (e.g., visible light, X-rays, microwaves, radio waves, infrared radiation, ultraviolet waves, cosmic rays, etc.).

Environmental Lapse Rate – The actual rate at which air temperature changes with height, typically as measured by a radiosonde.

Equilibrium Level (EL) – On a sounding, the level above the **Level of Free Convection (LFC)** where the temperature of a rising air parcel again equals the temperature of the environment. The height of the EL is the height where thunderstorm updrafts no longer accelerate upward. Thus, to a close approximation, it represents the expected height of the tops of thunderstorms.

Evaporation – The (isothermal) process by which a liquid is transformed into a gas.

Evapotranspiration – The loss of water from the soil both by evaporation and by **transpiration** from the plants growing thereon.

Eye – The roughly circular area of a hurricane marked by only light or completely calm winds with no precipitation and clear or lightly overcast skies.

F

Forcing – In meteorology, the influence of an external force or forces to generate vertical motion.

Forecast – A statement predicting an event will occur.

Forecast Accuracy – The extent to which the forecasts correspond to the values obtained from observations.

Forecast Bias – A measure of the correspondence between **forecasts** averaged over time and observations averaged over time.

Forecast Funnel – A weather forecasting technique that begins with the analyses of synoptic-scale features followed by successive analyses of smaller scale features.

Freezing Rain – Rain that falls as liquid water but freezes upon impact to form a coating of ice upon the ground or any object it contacts.

Frequency Distribution – A chart or table showing the number of times when an event occurred, categorized by discrete intervals.

Front – The boundary or transition zone between two dissimilar air masses.

G

Gaussian – Having a distribution that represents the probability that a continuous data-type (e.g., temperature) will be **symmetrical** about the mean and that it can be displayed graphically as a bell-shaped curve.

Geopotential Height – A measure of height scaled by the acceleration due to gravity at the surface (9.8 m s^{-2}). It is related to the potential energy that a 1-kilogram mass would have if it were lifted (against the pull of gravity) from sea-level to a given physical height. Geopotential height is used as the vertical coordinate for many meteorological applications. For most applications, geopotential height and the actual height above sea level (altitude) are interchangeable.

Geostrophic Wind – Winds that result from an exact balance of the **pressure gradient force** and the **Coriolis force**. On an isobaric chart, geostrophic winds are parallel to the height contours.

Global Forecast System (GFS) – A numerical forecast model operated by the National Centers for Environmental Prediction of the National Weather Service.

Gradient – In general, the spatial change of a physical quantity (e.g., temperature).

Gust Front – A boundary between cold air from the thunderstorm downdraft and warm, humid surface air.

H

Heat – A form of energy transferred between objects or systems as a result of a difference in temperature; not the same as temperature.

Heatburst – A thunderstorm downdraft characterized by a sudden, localized increase in air temperature near the ground.

Hypsometric Equation – An equation relating the thickness (i.e., vertical distance) between two pressure surfaces with the mean temperature of the layer.

I

Ice Pellets – Frozen precipitation which forms from the freezing of raindrops or the refreezing of partially melted snowflakes.

Ideal Gas Law (or Equation of State) – The state of an ideal gas can be shown to be p= RT, where p is the pressure; is the density; R is the specific gas constant; and T is the absolute temperature (in Kelvin).

Infrared Radiation – Radiation that is less energetic than visible radiation and more energetic than microwave radiation; the radiation emitted by the earth's surface or atmosphere. Also known as longwave radiation.

Infrared Satellite Image – An image taken from a satellite-based device that measures infrared wavelengths (e.g., 10.2 – 11.2 µm) emitted from the earth's surface.

Infrared Satellite Imagery – Satellite imagery that uses infrared wavelengths (e.g., 10.2 – 11.2 µm) to detect atmospheric phenomena during both the day and night. On an infrared image, each pixel is assigned a color according to the measured temperature. Typically, white indicates a very cold temperature and black indicates a warm temperature.

Insolation – Incoming solar radiation; sunshine. Also known as shortwave radiation.

Instability – The tendency for an object, if slightly moved, to accelerate in the direction of initial movement. In particular in meteorology, the tendency for air parcels to accelerate upward after being lifted.

Inversion – A layer with increasing temperature with increasing height; a reversal of the normal atmospheric temperature gradient with height in the troposphere.

Irradiance – The incident radiant flux (in watts per square meter [W m^{-2}]), received on a unit area.

Isobar – A line connecting points of equal pressure.

Isobaric – Characterized by equal or constant pressure with respect to either space or time.

Isobaric Charts – Weather maps that display meteorological data on a surface of constant pressure (e.g., a 700-mb chart).

Isodrosotherm – A line connecting points of equal dewpoint temperature.

Isopleth – A general term for a contour line connecting points of equal value of some quantity. Isobars and isotherms are examples of isopleths.

Isotach – A line connecting points of equal wind speed.

Isotherm – A line connecting points of equal temperature.

J

Jet Stream – Relatively strong winds concentrated within a narrow current in the atmosphere, normally referring to horizontal winds in the middle and upper troposphere.

K

K-Index – A common measure of the potential for non-severe thunderstorms. The K-Index can be computed using data from an atmospheric sounding.

Kelvin – A measure of temperature using an absolute temperature scale in which a change of 1 Kelvin equals a change of 1 degree Celsius; 0 K is the lowest temperature on the Kelvin scale. Water freezes at 273.15 K.

L

Landfall – The arrival of a hurricane onto land from its journey by sea.

Landfalling Hurricane – A hurricane that makes landfall at some time during its life span.

Lapse Rate – The rate at which air temperature changes with height.

Latent Heat – The heat released or absorbed by a substance during a phase change.

Left Mover – A thunderstorm that moves to the left relative to the main steering winds and to other nearby thunderstorms; often the northern part of a splitting storm.

Level of Free Convection (LFC) – The level in the atmosphere at which an air parcel that is lifted dry adiabatically until saturated and moist adiabatically thereafter would first become warmer than the surrounding air. Above the LFC, the parcel will rise freely until it becomes colder than the surrounding air.

Lifted Index (or LI) – A common measure of atmospheric instability. The lifted index provides an estimate of how unstable an air parcel would be if the parcel were lifted from the surface to 500 mb.

Lifting Condensation Level (LCL) – The level in the atmosphere at which an unsaturated air parcel lifted dry adiabatically becomes saturated.

Lifting Mechanism – Any method by which air is lifted.

Linear Regression – A statistical method used to predict the values of one dependent variable from known values of one or more independent variables using a linear mathematical equation.

Longwave Radiation – The infrared radiation emitted from the earth, having a wavelength greater than that of visible light; also called terrestrial radiation.

M

Maritime Polar Air Mass – An air mass characterized by cold, moist air.

Maritime Tropical Air Mass – An air mass characterized by warm, moist air.

Mature Stage (of a Thunderstorm) – The stage in the life cycle of a typical thunderstorm that occurs when the storm reaches its maximum height (e.g., 12 km or higher). The mature storm can produce heavy, convective precipitation, hail, frequent lightning, and strong winds.

Mean – The arithmetic average of a series of data, obtained by first summing all of the values and subsequently dividing by the number of observations.

Median – The middle value of a ranked (sorted) series of data.

Mesonet (or Mesonetwork) – A regional network of observing stations with station spacing (and frequency of observations) such that weather features on the mesoscale can be resolved.

Mesoscale – Of or relating to meteorological phenomena approximately 2 to 200 kilometers in horizontal extent; thunderstorms and squall lines are two examples of mesoscale events.

Meteogram – A graphical representation of how meteorological variables (temperature, solar radiation, etc.) change with time. The meteogram can be constructed using observed data or forecast data.

Microburst – An intense downdraft less than 4 km wide (about 2.5 miles) that may occur beneath a thunderstorm.

Mixing Ratio (w) – In an air parcel, the mixing ratio is the ratio of the mass of water vapor to the mass of dry air.

Mode – The value(s) of a data series that occurs with greatest frequency.

Model Output Statistics (MOS) – In numerical weather prediction, statistical relationship between forecast variables and observed weather variables, typically used for prediction of variables not forecast by the model.

Moist Adiabat – On a thermodynamic diagram, a line representing the temperature change with height for a moist (saturated) adiabatic process.

Moist Adiabatic Lapse Rate – The rate at which air temperature changes with height for a saturated parcel of air.

Moisture Tongue – An area of relatively high dew point values that can be traced to a body of water; also known as a moisture ridge.

Multi-Cell Thunderstorm – A thunderstorm consisting of two or more cells, of which most or all are often visible at a given time as distinct domes, or cloud towers, in various stages of development.

N

Nested Grid Model (NGM) – One of the operational forecast models run at the National Centers for Environmental Prediction of the National Weather Service. The NGM is run twice daily, with forecast output out to 48 hours.

NEXRAD (Next-Generation Weather Radar) – The network of high-resolution Doppler radars operated by the National Weather Service; NEXRAD units are known as WSR-88D.

Normal – The average value of a weather element (e.g., temperature, precipitation, humidity) during a uniform and relatively long interval (e.g., 30-year period). Normal assumes a Gaussian distribution, whereby the central tendency of the distribution is given as the normal value.

Numerical Weather Prediction (NWP) – The forecasting of the evolution of atmospheric disturbances by computational methods on a computer.

O

Onshore – Coming from the sea toward the land.

Orographic – Of or relating to mountains or hills.

Orographic Lift – The lifting of air caused by its passage up and over mountains or other sloping terrain.

Orography – The nature of an area with respect to elevated terrain, including mountains.

Outflow – Rain-cooled air moving outward from beneath the base of a thunderstorm; it occurs when a thunderstorm downdraft strikes the ground and spreads outward.

P

Power – The rate at which energy is transmitted or exchanged, often expressed in units of watts, where 1 watt = 1 joule per second.

Precipitation Mode – A less sensitive operational mode of NEXRAD radar. The radar automatically switches into precipitation mode from the more sensitive clear-air mode if the measured base reflectivity exceeds a specific threshold value. In precipitation mode, the radar obtains only 5 elevation slices in 4-6 minutes. These echoes consist of raindrops, sleet, and hailstones.

Pressure Gradient Force – The force resulting from a change in pressure over a given distance at a given time.

Prevailing Wind Direction – The wind direction most commonly observed during a given period.

Probability Distribution – A mathematical description that represents the chance that a given event will occur, displayed across a range of possible values.

Probability Distribution Function – A function that yields the probability that a random variable will assume some value.

Probability of Exceedence – The likelihood that an observation will be greater than or equal to a selected value.

Probability of Occurrence – The likelihood that an observation will be equal to a selected value or within a range of values.

Probability of Precipitation (POP) – As defined and used by the National Weather Service, the likelihood, expressed as a percent, of a measurable precipitation event (1/100th of an inch) at a model grid point during the indicated valid period.

Pyranometer – A type of radiation sensor that measures the combined intensity of incoming direct solar radiation and the diffuse sky radiation.

R

RADAR – An acronym for <u>ra</u>dio <u>d</u>etection <u>an</u>d <u>r</u>anging; a radio device or system for locating an object by means of ultrahigh-frequency radio waves reflected from the object and received, observed, and analyzed by the receiving part of the device in such a way that characteristics (e.g., distance and direction) of the object may be determined.

Radiation – (1) The process by which radiated energy moves through space or a material media; (2) Energy propagated through space or through a material media in the form of an advancing disturbance in electric and magnetic fields (e.g., visible light, X-rays, microwaves, radio waves, infrared radiation, ultraviolet waves, cosmic rays, etc.). Radiation can be absorbed, reflected, scattered, refracted, or transmitted.

Radiation Inversion – A shallow layer of air near the surface where the temperature increases with height. It is caused by radiational cooling at the earth's surface during the night.

Radiosonde – A meteorological instrument package and miniature radio transmitter that are carried aloft by an unmanned balloon for the simultaneous measurement and transmission of meteorological data.

Rainband – The structure of both cloud and precipitation associated with an area of rainfall that is elongated sufficiently to detect and assign an orientation (e.g., north-south).

Range – A measure of variation of a data series, determined by subtracting the lowest value of the series from the highest value.

Reflection – The process whereby radiation (or other waves) incident upon a surface is directed back into the medium through which it traveled.

Refraction – The process whereby radiation experiences a change in direction as a result of a change in density of the medium or media through which it travels.

Relative Humidity – A measure of the water vapor content of the air at a given temperature; the ratio of the actual amount of moisture in the air to the maximum amount that the air could contain at the same temperature (e.g., actual vapor pressure to saturation vapor pressure or the actual mixing ratio to the saturation mixing ratio). It is expressed as a percentage.

Return Period – A recurrence interval used in frequency analysis to represent the average time interval between the occurrence of a given quantity and that of an equal or greater quantity. Used to delineate the 100-year floodplain, for example.

Ridge – An elongated area of relatively high atmospheric pressure; the opposite of a trough.

Right Mover – A thunderstorm that moves appreciably to the right relative to the main steering winds and to other nearby thunderstorms. Right movers typically are associated with a high potential for severe weather.

S

Saffir-Simpson Hurricane Scale – A category system developed in the 1970s that describes the intensity of a hurricane based on a combination of factors, including wind speed, rainfall, and height of the storm surge.

Satellite Image – A two-dimensional representation of the reflectance values measured by a satellite.

Satellite Imagery – Photography of the earth taken from devices on board satellites. Satellite imagery provide information about cloud cover, moisture content, temperature, winds, and other atmospheric constituents.

Saturation (of air) – The atmospheric condition whereby the amount of water vapor in the atmosphere is the maximum possible that can exist at a given pressure and temperature.

Saturation Mixing Ratio – The ratio of the mass of water vapor to the mass of dry in a saturated parcel; the maximum mixing ratio an air parcel can have at a given temperature.

Saturation Vapor Pressure – The vapor pressure of a system in which vapor and liquid (or vapor, liquid, and solid) coexist in equilibrium (i.e., where the air can contain no more vapor molecules at that temperature).

Scalar – A physical property or quantity that is described by a single numerical value at each point in space. A scalar has no direction, just a magnitude. In meteorology, common scalar quantities are temperature, pressure, and humidity.

Scalar Field – A spatial distribution of a **scalar** variable (such as temperature or moisture).

Scattering – The process by which small particles, including photons in a beam of electromagnetic radiation, are forced to change their direction of motion.

Sensible Heat – The heat absorbed or transmitted when the temperature of a substance changes but the substance does not change phase.

Sequence – A series of numbers in a specified order (e.g., maximum temperatures in order of calendar day).

Series – A number of related items (e.g., rainfall totals) listed one after another.

Severe Thunderstorm – A thunderstorm with wind gusts of 50 knots (58 mph) or greater, hail at least three-quarters of an inch in diameter, or a tornado or funnel cloud.

Shortwave Radiation – In meteorology, radiation having a wavelength equal to or less than that of visible light, typically from the sun. Sometimes called solar radiation.

Single-Cell Thunderstorm – A thunderstorm consisting of only one updraft/downdraft couplet.

Sleet – A mixture of rain or snow with ice pellets.

Solar Constant – The radiative energy from the sun as it strikes an imaginary surface perpendicular to the sun's rays at the top of the atmosphere; measured when the earth is at its mean distance from the sun.

Solar Noon – The time when the sun reaches its highest point in the sky.

Solar Radiation – The radiation emitted by the sun.

Sounding – A plot of the vertical profile of temperature, dew point, and winds above a fixed location, usually constructed from radiosonde data; used extensively in weather forecasting.

Sounding-based Stability Index – An index calculated from balloon observations (e.g., lifted index, K-Index, Total-Totals Index, SWEAT Index) that provides guidance about the potential organization, type, and severity of thunderstorms (e.g., supercell, multi-cell).

Specific Heat – The amount of heat required to raise the temperature of 1 gram of a substance by 1 degree Celsius.

Splitting Storm – A thunderstorm that splits into two storms that follow diverging paths (i.e., a **left mover** and a **right mover**). The left mover typically moves faster than the original storm, the right mover, slower. Of the two, the left mover is most likely to weaken and dissipate (but on rare occasions can become a very severe storm), while the right mover is the one most likely to become a supercell.

Squall Line – Any line or narrow band of active thunderstorms that is not directly along a frontal boundary.

Stability – The atmospheric state in which an air parcel that is pushed either upward or downward will tend to return to the location from where it was pushed.

Stability Indices – Quantities that are derived from sounding data and are used to evaluate the potential for convection, including severe convection. Stability indices include the lifted index and SWEAT index.

Stable (Air) – Generally, air that returns to its initial position when forced to move vertically.

Standard Deviation – A measure of the variance, or spread, of values about the mean of a data series. One standard deviation contains approximately 68% of the values in the series, and twice the standard deviation contains approximately 95% of the values in the series.

Station Model – The specific pattern for plotting meteorological data and symbols on a weather map that describes the state of the weather at that geographic location.

Statistic – A numerical function or value that describes a sample or population of data.

Stefan-Boltzmann Law – A mathematical relationship for electromagnetic radiation that states the irradiance of a blackbody is proportional to the fourth power of the absolute temperature (in Kelvin) of the blackbody.

Storm Scale – Referring to weather systems with sizes on the order of individual thunderstorms.

Storm Surge – An atypical rise of the sea along a shore primarily resulting from the winds of a storm, especially those of a hurricane.

Straight-Line Winds – Generally, any wind that is not associated with rotation, used mainly to differentiate them from tornadic winds.

Stratiform Precipitation – Precipitation that does not result from convection; typically, precipitation from a stratus cloud deck.

Stratus – A low-level cloud in the form of a gray layer with a rather uniform base.

Summer Solstice – The solstice when the sun at solar noon is highest in the sky, marking the first day of summer (~June 21). On this day, the sun is directly over the Tropic of Cancer at solar noon in the Northern Hemisphere.

Supercell Thunderstorm – A violent thunderstorm containing updrafts and downdrafts that are nearly in balance, allowing it to maintain itself for several hours. Supercells can produce hail and large tornadoes. Supercell thunderstorms typically rotate.

Supercooled Liquid Water – Water that exists in its liquid phase at temperatures below 0°C.

Surface Layer – A shallow layer of air (<100 m thick, in general) next to the ground where wind-generated turbulence (e.g., from friction) exceeds convectively generated turbulence (e.g., from surface heating).

Surface Map – A map or chart showing the principal meteorological elements near the earth's surface at a given time and over an extended region.

SWEAT Index – The Severe Weather Threat (SWEAT) Index provides a measure of instability, low-level moisture, wind speed, and wind shear. It can be calculated directly from sounding data and serves as an important tool for determining both severe thunderstorm potential and tornado potential.

Symmetric – Having balanced proportions.

Synoptic Map – A map that displays weather conditions, as they exist simultaneously over a broad area.

Synoptic Scale – The horizontal scale of the migratory high and low pressure systems of the lower troposphere; generally considered 1000 to 2500 km in length.

T

Temperature – The degree of hotness or coldness as measured on some definite scale (e.g., Fahrenheit, Celsius, or Kelvin). Temperature is proportional to the motion of the molecules of a substance (i.e., more molecular motion leads to higher temperatures).

Terrestrial Radiation – The total amount of infrared radiation that is emitted from the earth's surface or atmosphere. Also known as longwave radiation.

Thermal Ridge – An area of relatively high values of temperature.

Thermodynamic Diagram – A chart containing isopleths (contours) of pressure, temperature, moisture, and potential temperature drawn relative to each other such that basic thermodynamic laws are satisfied (e.g., conservation of energy). Such a chart typically is used to plot atmospheric soundings, and to estimate potential changes in temperature, moisture, etc., if air were displaced vertically from a given level. A thermodynamic chart thus is a useful tool in diagnosing atmospheric instability.

Thickness – The vertical depth between two different surfaces of atmospheric pressure.

Thin Line – A line of base reflectivity that separates air of differing densities. Thin lines typically are located ahead of thunderstorms (i.e., associated with the gust front), but can be associated with cold fronts and dry lines. The thin line is caused by backscattering of the radar's pulse from a gradient in the index of refraction of air (i.e., a sharp contrast in air density across a short distance).

Thunderstorm Outflow – A current of typically cool, moist air from a thunderstorm downdraft that strikes the ground and spreads radially outward.

Topography – The shape and composition of the surface landscape, including both natural and manmade features. Topographical features include the distribution of mountains, valleys, and human settlements and the patterns of rivers, roads, and railways.

Tornado – A violently rotating column of air protruding from a cumulonimbus cloud (i.e., thunderstorm) and in contact with the ground.

Total–Totals Index – A measure of stability and used as a severe weather forecast tool. This index has proven to be useful in diagnosing regions of severe thunderstorms.

Towering Cumulus – A large cumulus cloud with great vertical development, usually with a cauliflower-like appearance, but lacking the characteristic anvil of a cumulonimbus.

Transmission – The movement of incident radiation completely through a medium (e.g., the air).

Transpiration – The process by which plants transfer stored water to water vapor in the atmosphere.

Trend – A general tendency toward a particular direction (e.g., increasing temperatures).

Trendline – A line that represents the trend of a sequence of data points, typically computed using linear regression.

Tropical Cyclone – The general term for a large low-pressure system that originates over the tropical oceans; includes tropical depressions, tropical storms, hurricanes, and typhoons.

Tropical Storm – A **tropical cyclone** with winds stronger than 27 knots but less than 66 knots.

Tropopause – The upper boundary of the troposphere, usually characterized by an abrupt change in how the temperature changes with height. Below the tropopause, temperature generally decreases with height; above the tropopause, temperature generally increases with height.

Troposphere – The portion of the atmosphere that extends outward about 10 to 20 km from the earth's surface, and in which generally temperature decreases rapidly with altitude, clouds form, and convection is active.

Trough – An elongated area of relatively low atmospheric pressure; the opposite of a ridge.

U

Unstable (Air) – Generally, air that will continue to rise and accelerate when briefly forced upward.

Updraft – A relatively small-scale current of air with marked upward motion.

Upwelling Radiation – The component of radiation (either reflected solar or emitted terrestrial) directed upward from the earth's surface.

V

Vapor Pressure – In meteorology, the pressure exerted only by molecules of water vapor in the air.

Variance – A measure of the variability, or spread, in a dataset; equal to the square of the standard deviation.

Vector – A physical property or quantity that is described by both a magnitude and direction at each point in space. In meteorology, common vector quantities include wind velocity and storm motion.

Vernal Equinox – The equinox when the sun reaches a point directly above the equator at solar noon on its northward journey over the Northern Hemisphere. It marks the start of the astronomical spring in the Northern Hemisphere (~21 March).

Vertical Temperature Profile – A graphical representation of the air temperature with increasing altitude or decreasing pressure.

Vertical Wind Profile – A graphical representation of the wind speed and direction with increasing altitude or decreasing pressure.

Virga – Water or ice particles falling from a cloud but evaporating before reaching the earth's surface.

Visible Satellite Image – An image taken from a satellite-based device that measures the visible wavelengths reflected from the earth's surface.

Visible Satellite Imagery – Satellite imagery that uses visible wavelengths to detect atmospheric phenomena during daylight. Visible satellite images provide information about cloud cover. Typically, areas of white indicate clouds and shades of gray indicate clear skies. Thicker clouds appear whiter than thinner clouds on a visible image. Visible imagery is equivalent to what a human would see if onboard a spacecraft.

W

Warm Front – The advancing edge of a warm air mass.

Water Vapor Image – An image taken from a satellite-based device that measures infrared wavelengths from 6 to 7 μm to detect atmospheric water vapor in the middle and upper troposphere.

Water Vapor Imagery – Satellite imagery that uses infrared wavelengths from 6 to 7 μm to detect atmospheric water vapor. When viewing a water vapor image, typically white indicates a deep moist layer or a cloud in the upper troposphere, and black indicates radiation from the earth's surface or a dry layer in the middle and upper troposphere.

Watt – The derived unit for power. One watt is equal to one joule per second. Named for James Watt (1736-1819), a Scottish engineer.

Weather – The state of the atmosphere with respect to various weather elements. Weather is thought of in terms of temperature, humidity, precipitation, cloudiness, visibility, and wind, as well as tornadoes, thunderstorms, hurricanes, and other natural atmospheric phenomena.

Wein's Displacement Law – A mathematical relationship for electromagnetic radiation that states the wavelength of maximum radiation (i.e., peak intensity) for a blackbody is inversely related to its absolute temperature (in Kelvin).

Wet-Bulb Temperature – The lowest temperature that can be obtained by evaporating water into the air at constant pressure. The name comes from the technique of putting a wet cloth over the bulb of a mercury thermometer and then blowing air over the cloth until the water evaporates. Because evaporation requires heat, the thermometer cools to a lower temperature than a thermometer with a dry bulb at the same time and place. Wet bulb temperatures can be used along with the **dry-bulb temperature** to calculate dew point or relative humidity.

Wind Barb – A graphical representation of wind speed and direction that is used on weather maps. Wind barbs point in the direction from which the wind blows and graphically indicate the wind speed with a combination of flags and pennants.

Wind Direction – The direction from which the wind is blowing.

Wind Shear – The local variation of the wind speed or wind direction. Wind shear usually refers to vertical wind shear (i.e., the change in wind speed or direction with height).

Wind Speed – The magnitude of the wind velocity; the ratio of the distance traveled by the air to the time taken to cover the distance.

Windward – Facing the direction from which the wind is blowing. Often refers to the prevailing wind direction (e.g., the windward side of the Rocky Mountains is the west side).

Winter Solstice – The solstice when the sun at solar noon is lowest in the sky, marking the first day of winter (~December 21). On this day, the sun is directly over the Tropic of Capricorn at solar noon in the Northern Hemisphere.